Immanent Externalities

# Historical Materialism Book Series

The Historical Materialism Book Series is a major publishing initiative of the radical left. The capitalist crisis of the twenty-first century has been met by a resurgence of interest in critical Marxist theory. At the same time, the publishing institutions committed to Marxism have contracted markedly since the high point of the 1970s. The Historical Materialism Book Series is dedicated to addressing this situation by making available important works of Marxist theory. The aim of the series is to publish important theoretical contributions as the basis for vigorous intellectual debate and exchange on the left.

The peer-reviewed series publishes original monographs, translated texts, and reprints of classics across the bounds of academic disciplinary agendas and across the divisions of the left. The series is particularly concerned to encourage the internationalization of Marxist debate and aims to translate significant studies from beyond the English-speaking world.

*For a full list of titles in the Historical Materialism Book Series available in paperback from Haymarket Books, visit:* www.haymarketbooks.org/series_collections/1-historical-materialism.

# Immanent Externalities

*The Reproduction of Life in Capital*

Rebecca Carson

Haymarket Books
Chicago, IL

First published in 2023 by Brill Academic Publishers, The Netherlands
© 2023 Koninklijke Brill NV, Leiden, The Netherlands

Published in paperback in 2024 by
Haymarket Books
P.O. Box 180165
Chicago, IL 60618
773-583-7884
www.haymarketbooks.org

ISBN: 979-8-88890-215-8

Distributed to the trade in the US through Consortium Book Sales and Distribution (www.cbsd.com) and internationally through Ingram Publisher Services International (www.ingramcontent.com).

This book was published with the generous support of Lannan Foundation, Wallace Action Fund, and the Marguerite Casey Foundation.

Special discounts are available for bulk purchases by organizations and institutions. Please call 773-583-7884 or email info@haymarketbooks.org for more information.

Cover art and design by David Mabb. Cover art is a detail of *Duvet/Ground,* acrylic paint on machine stitched Morris & Co. duvet cover (2003).

Printed in the United States.

Library of Congress Cataloging-in-Publication data is available.

# Contents

Acknowledgements IX

Introduction 1

1 Fictitious Capital and the Re-emergence of Personal Forms of Domination 12
    Introduction 12
  1.1 Fictitious Capital 16
  1.2 Fictitious Capital and Value Form 22
  1.3 Social Reproduction and Personal Domination 28

2 Money Form 33
    Introduction 33
  2.1 Political Subjectivity and the Monetary Link between Italian Operaismo and Capital Logic 37
    2.1.1 *1974 Monetary Crisis* 37
    2.1.2 *New Laws for Action* 38
    2.1.3 *Classe Operaia Literature* 44
    2.1.4 *Money and Financialisation* 46
    2.1.5 *Money as Capital* 48
    2.1.6 *Capital Logic Critique* 52
  2.2 Money as Money 59
    2.2.1 *The Genesis of Money* 59
    2.2.2 *Money as Medium of Reproduction* 61
    2.1.3 *Money as Social Form* 62
    2.2.4 *Money's Externality* 68

3 Fetish Character 71
    Introduction 71
  3.1 The Presupposition of Reification and The Money Form 73
    3.1.1 *The Hegelian Movement of Value as Essence* 76
    3.1.2 *Value as Essence in Labour Time* 79
    3.1.3 *The Driving Force of the Process* 80
    3.1.4 *The Fetish Character of Money* 81
    3.1.5 *The Automatic Fetish of Interest-Bearing Capital* 85
  3.2 Personal and Impersonal Forms of Domination 90
    3.2.1 *Marx's Use of the Category Person* 90
    3.2.2 *Person as Juridical Mask* 93

      3.2.3   *Interpersonal Relations*   94
      3.2.4   *Extra-Economic Forms of Domination*   101

## 4 Time and Schemas of Reproduction   104
    Introduction   104
  4.1   The Circulation of Capital   108
  4.2   Interruptions and Differential Temporal Forms within Capital's Reproduction   113
  4.3   Marx's Presentation of The Metamorphoses of Capital and Their Circuit   116
  4.4   Marx's Presentation of The Turnover of Capital   121
  4.5   Marx's Presentation of the Reproduction and Circulation of Total Social Capital   122
  4.6   The Three Circuits of Capital   127
  4.7   The Role of the Credit System within Capital's Reproduction   128
  4.8   Expanded Reproduction   130
  4.9   A Complete Concept of Money for Understanding Capital's Reproduction   133
  4.10  Non-capitalist Variables within Capital's Reproduction   136
    Conclusion   138

## 5 Marx's Social Theory of Reproduction   140
    Introduction   140
  5.1   Capital's Life Process   149
  5.2   Intersubjective Structures   159
  5.3   The Category of Reproduction in Hegel's The Science of Logic   164
  5.4   Concrete Reproduction of Human Life and Nature   172
  5.5   Marx's Two Concepts of Life   182
    Conclusion   184

**Bibliography**   189
**Index**   199

# Acknowledgements

An enormous beholden thank you to Étienne Balibar who accompanied me throughout this project with patient adherence and extraordinary pedagogical generosity. Thank you to Jillian Carson and David Carson for their grounding constancy and to everyone at the CRMEP who together formed a uniquely cultivating intellectual community. A special thanks to Christopher J. Arthur, Dave Beech, Riccardo Bellofiore, Maria Chehonadskih, Alex Fletcher, Peter Hallward, Peter Osborne, Flo Ray, Eric-John Russell and Massimiliano Tomba who read drafts, and provided imperative feedback.

This book could not have been possible without the unwavering fervent support of Thomas Jungnitz Watson, to whom it is dedicated with abounding gratitude and love.

Earlier versions of chapters, now much changed, have appeared in Continental Thought & Theory: A Journal for Intellectual Freedom (2017), Consecutio Rerum (2018), The Politics of the Many: Contemporary Radical Thought and the Crisis of Agency (2022) and Critical Theory Today: On the Limits and Relevance of an Intellectual Tradition (2022).

# Introduction

Marx's *Capital* is an unfinished project.[1] Where opacity and indeterminacy persist across its three volumes, methodological reconstruction is required. This book philosophically reconstructs Marx's critique of political economy from the perspective of reproduction. Such a reconstruction is pursued by examining the process of reproduction in *Capital* through the lens of a value analysis.[2] A value analysis is deployed to address significant, overlooked form determinations underpinning the reproduction of life in *Capital* the book. This task is pressing not only because of *Capital's* incompletion, but because the ever-changing relationality between the logic of value accumulation and the concrete world obliges fuller analysis.

The reconstruction ventured here develops from the claim that capital's reproduction process entails a central contradiction between capital's abstractions and non-capitalist life-making processes. This can be logically derived from the primacy of the money form in Marx's analysis as the medium of reproduction: 'money as money'. 'Money as money', as a medium of reproduction of capital's value forms, is both capitalist and non-capitalist.[3] Thus, this book

---

1 Many esteemed readers of Marx have made this claim from differing perspectives. Notably, Étienne Balibar in his book, *The Philosophy of Marx*, claims that *Capital* has several incomplete philosophical projects. Balibar further contends that reading Marx as a philosopher not only positions one 'at a remove from doctrine' but that objectively, through textual evidence of what Marx produced and was able to produce, 'that doctrine does not exist'. Another notable reader of Marx, Michael Heinrich, has made claim to the incomplete nature of *Capital* in his *An Introduction to the Three Volumes of Karl Marx's Capital*, albeit from a political perspective. Heinrich insists that no critique of bourgeois relations can be complete without a theory of the state. Meanwhile, from the perspective of money and credit, Alex Callinicos has claimed that although these elements are developed, they are done so incompletely in Volume III. In his book *Karl Marx's Ecosocialism: Capital, Nature, and the Unfinished Critique of Political Economy*, Kohei Saito insists that without full address of the natural sciences, including physiology and ecology, Marx's critique remains unfinished. This is evidenced through Marx's partial manuscripts that detail the natural sciences as imperative to his critique. See Balibar 2014; Callinicos 2021; Heinrich 2012; Saito 2017.

2 This entails a commitment to reading *Capital* as a book that exposits an ontology that depends on the formal role of value and its circulation within capital's social relations. Readers of Marx, working broadly in this way include I.I. Rubin, Michael Heinrich, Hans-Georg Backhaus, Christopher J. Arthur and Diane Elson. For a detailed explanation of the German tradition behind this reading of Marx, see Bellofiore and Redolfi Riva 2015, pp. 24–36.

3 Rather than opposing the 'economic' with the 'extra-economic' – where the latter is seen as that which provides the conditions for the reproduction of capital – reproduction relies

contends that capital's conditions of reproduction must be other to capital while also, counterintuitively, posited by capital as its presupposition. This contention has repercussions for how one understands the lived effects of historical development, such as when the proportion of credit money in circulation increases, where capital's reproduction and accumulation of wealth is not predominantly governed by the accumulation of capital or 'valorised value'. This theoretical pursuit has practical political meaning because – if analysis can locate non-capitalist practices and processes internal to capital – new arenas of struggles over social reproduction open up to contestation.

The expansion of finance capital forms the backdrop to this book's presentation. Rather than a new stage in capital's development – eroding the primacy of production – this book positions finance's extension as a marker of the increased circulation of fictitious capital[4] within a globally shifting – yet nonetheless strong – production process. Financialisation is a growing asymmetry between circulation and production. Accordingly, barriers to the reproduction of capitalist production surfacing as crisis ought not to be attributed to a breakdown of capitalism as a social form. Rather, financialised forms of generating profit are incurred through the circulation of money that does not move through the production process. In Marxist terms, this is technically unvalorised value whose profit is 'fictitious'. Far from contributing to the disintegration of capitalist social relations (that depend on the production process as one aspect of a larger structure of relations), the increased circulation of fictitious capital allows for the devaluation of the wage and state austerity while at the same time keeping capitalist relations formally afloat. Consequently, life-sustaining practices of care are rendered increasingly difficult for the majority of those living in capitalist societies. This dynamic – manifest as a strain on social reproduction – is not a strain on the reproduction of the capitalist relation. It is neither a disintegration nor a deepening of capital's abstract form. Instead, it reflects a shift in modes of subjection, or structural conditions of life-making processes, within capitalist relations due to changes in the constitution of money in circulation.

Considering these conditions, this book argues that the fundamental contradiction of capital resides in capital's reproduction, a process that necessarily relies on non-capitalist elements and relationships. Capital's self-reproduction

---

    on economic forms of reproduction that are non-capitalist as well as extra-economic non-capitalist forms.

4  Fictitious capital is Marx's term for credit money circulating through financial operations and is discussed in *Capital* Volume III (Marx 1991).

requires the resistant heterogeneity of 'immanent externalities':[5] non-capitalist practices and processes, produced within capitalist social relations, necessary for the continuation of capitalist exploitation and accumulation, and therefore also necessary for the reproduction of the totality of social capital. Non-capitalist elements and processes remain non-capitalist due to their irreducibility to capital's abstract form and not because they belong to a distinct social system. This irreducibility obtains both abstractly, within the formal movement of value itself (where money, as capital's medium of reproduction, is both capitalist and non-capitalist), and concretely, within the reproduction of the objectivity of human life and nature (where organic life engenders moments of interruption from capital's self-movement). The role of immanent externalities becomes clear when surveying all three volumes of *Capital*, in which Marx deploys two distinct concepts of 'life': that of capital's life on the one hand and that of organic matter on the other. Marx ascribes capital a life-like manner due to its self-reproducing nature as automatic subject, where its medium of reproduction is the money form, which is both capitalist and non-capitalist. By contrast, Marx then ascribes human actors – who are often personifications of the capitalist relation, such as property owners or labourers – an irreducibly non-capitalist, organic life, existing in metabolistic relation to nature. This methodological contradiction forms the decisive argument of this book.

The set of concerns introduced so far, concerning life and reproduction, attest to latent difficulties in Marx's work. The resolution of such difficulties is relevant to urgent political and analytical problems, such as how one might account for interpersonal tensions and exploitations, like those of race and gender, within the context of an impersonal economic system. While the accumulation of capital is based on impersonal forms of subjection, which constitute capital's fetish character, interpersonal tensions are greatly exacerbated by the forms of subjection underpinning the growing use and circulation of fictitious capital (because fictitious capital depends on interpersonal juridical relations). This exacerbation of interpersonal antagonisms occurs within the context of capital's self-movement, a process traditionally characterised by impersonal subjection. Hence, subjection to capital's forms does not only take place within the sphere of production but also within the sphere of circulation, where personal relations uphold the formal character of capital accumulation.[6]

---

5  This term is derived from Suzanne de Brunhoff's address of the money form when describing its nature as a medium of reproduction in circulation (de Brunhoff 1978, p. 37).
6  While abstract labour is a condition for value, value is only valorised because of its circulation, which works to reproduce capital's social form as a process of valorisation. Within this process, interpersonal forms of subjection are not a negation of negation of fetishism (revers-

By analysing the link between fictitious capital and social reproduction, this book demonstrates how the increased circulation of fictitious capital leads to an increased stress on interpersonal forms of domination, or non-capitalist relations, which are crucial for successful capitalist reproduction. The expanding circulation of fictitious capital is a deepening political concern today, where barriers to the reproduction of individual life have grown through economic polarisation and ecological degradations. Because fictitious capital is never valorised value, it not only fails to distribute a wage to working people but, at the level of state debt, provides rationale to defund social resources. In the wealthier states, as debt has mounted, state services have also suffered diminution. This dynamic has degraded resources for social reproduction – or life sustaining practices – in housing and healthcare, nourishment and child rearing. As such, non-state and non-capitalist actors have been required to compensate. Thus, the reproduction of capitalist relations increasingly requires its own opposite; what is other to the system upholds the life-like process of accumulation and automatic growth. In this regard, capital's externalities, or non-capitalist variables and subject positions, have borne greater stress when barriers to valorisation have surfaced.

The primary tension in capitalism is that between the reproduction of capital and the reproduction of human life and nature. By drawing on formal accounts of capital's reproduction – in which fictitious capital remains central – I argue that an adequate analysis of the above tension requires the analysis of two concepts of life derived from Marx's three volumes of *Capital*: one concrete and one abstract. The concrete concept refers to human life and nature, the abstract to the life of capital. Concrete life engenders natural limits that in turn create limits for capital's movement of value. Correspondingly, capital's abstractions determine the practical form of concrete life. The tension between these two forms of life lies at the core of this book. The book dialectically analyses what extent concrete life is produced by capital's abstract form and to what extent it remains independent. By methodologically – and philosophically – clarifying the dynamics of life and reproduction, the book begins the process of asking how concrete life can overcome its appropriation to the reproduction of capital's abstract social form.

To do this, this book begins with the chapter 'Fictitious Capital and the Re-emergence of Personal Forms of Domination'. This first chapter is propositional, reflecting a thesis that the book elaborates throughout. The components

---

ing the inversion of personal relations and thing-like relations) but a redoubling of the fetish through the deployment of interpersonal subjection within the formal dynamic of the fetish.

of its argument, deployed in each subsequent chapter, are axiomatic, reflecting the multiple elements necessary for its substantiation. The first chapter proposes that the re-emergence of interpersonal forms of domination[7] takes place in the context of the impersonal domination that structures the fetish character of capitalist social relations.[8] Furthermore, it argues, the increased circulation of fictitious capital underpins this re-emergence of personal forms of domination. When dominated by the circulation of fictitious capital, inhabitants of capitalist society become subject to interpersonal domination and dependencies existing within the sphere of the reproduction and circulation of capital. These dependences are constituted by social relationships that, while produced by capitalist relations, are not strictly capitalist, such as debtor-creditor contracts or kinship relations. Consequently, the chapter elaborates, social practices that facilitate the reproduction of human life bear greater stress when capital incurs barriers to valorisation engendered by interruptions in production.

Chapter 1 contends that fictitious capital is substantiated by future labour rather than past labour. This temporal discordance places increased pressure on social reproduction. This is because the social reproduction of labour is what insures the reproduction of future labour power. The variable of social reproduction, therefore, lies at the heart of fictitious capital's circulation, being required to reproduce both labour in the present and labour for the future. The dual demand imposed by fictitious capital hence subjects social reproduction to greater strain. By developing this proposition, the first chapter responds to literature that currently looks at the issues of financialisation and social reproduction separately or without a clear exposition of their co-determinacy. I formulate the logical connection between the two, showing that they cannot be treated as separate theoretical issues. A rigorous Marxian account of the interaction between financialisation and social reproduction has yet to develop at either conjunctural or structural levels. Recently, of course, much literature has addressed finance and financialisation from a Marxian register, including Costas Lapavitsas's *Profiting Without Producing: How Finance Exploits Us All*,[9] François Chesnais's *Finance Capital Today: Corporations and Banks in the Lasting*

---

7   Such as legal-, race- and gender-based interpersonal relations that are not directly mediated by capital's abstract forms.
8   Impersonal domination, or domination by capital's abstract form – constituting its fetish character – is the result of capital's valorisation or, in other words, of capital's accumulation in circulation and appearance in the form of 'money as capital', which relies on abstract labour subsumed through the production process.
9   Lapavitsas 2013.

*Global Slump*,[10] Michael Roberts's *The Long Depression: Marxism and the Global Crisis of Capitalism*,[11] and Cédric Durand's *Fictitious Capital: How Finance Is Appropriating Our Future*.[12] Similarly, there has been a revival of social reproduction theory within Marxist feminism, its notable works including Susan Ferguson's *Women and Work: Feminism, Labour, and Social Reproduction*,[13] Tithi Bhattacharya's edited collection *Social Reproduction Theory: Remapping Class, Recentering Oppression*,[14] Martha E. Gimenez's *Marx, Women, and Capitalist Social Reproduction*,[15] and Cinzia Arruzza's essay 'Remarks on Gender',[16] to name only a few. However, thus far, analysis has seldom delved into the logical co-production of finance capital *and* social reproduction. One salutary exception can be found in Nancy Fraser's work.[17] Yet Fraser's framework deviates from a capital logic approach, drawing instead on Polanyian economics. The philosophical explanation for the interrelation of finance and social reproduction – which is one of 'life' and 'form' – thus remains ambiguous in her work. By drawing on a value theoretical analysis, Chapter 1 looks to remedy a conspicuous methodological lacuna in Marxist accounts of the present.

Chapter 2, 'Money Form', unfolds in a methodological register distinct from the other chapters. This chapter provides an intellectual history of the concept 'money as money', which is the medium of capital's reproduction. This intellectual history is deployed to explore the institutional background out of which the concept grew. The chapter examines the monetary understanding of capital developed in a series of essays expounding a confrontation between members of the editorial collective of the Italian workerist journal *Primo Maggio* and French Marxist Suzanne de Brunhoff. Although untranslated original documents of the essays have appeared in self-published journals – some of which are reproduced in de Brunhoff's *The State, Capital and Economic Policy*[18] – there has been no significant commentary on this debate and its consequences for Marxian monetary theory. Consideration of this history is required to differentiate 'money as money' from proximate ideas in the Italian tradition, which, given the lack of secondary literature, is needed to delineate the very nature

---

10  Chesnais 2016.
11  Roberts 2016.
12  Durand 2017.
13  Ferguson 2020.
14  Bhattacharya 2017.
15  Gimenez 2018.
16  Arruzza 2014.
17  Exemplary of this is her series of essays for the *New Left Review*, including 'Behind Marx's Hidden Abode', 'Contradictions of Capital and Care' and 'Climates of Capital', Fraser 2021.
18  de Brunhoff 1978.

of the term. The chapter thus examines competing ideas and paths not taken, reflecting historical changes in the circulation of the money form.

Theoretically, the debate between these sides has made it possible to reflect upon the contradictory developments of Marxian value theory within the post-Bretton Woods world. The position internal to the *Primo Maggio* group, often classified as post-Marxist, conflicted with that attributed to de Brunhoff's. Presently, Marxism and post-Marxism represent two discrete positions attributed to different understandings of capitalism and the function of money in political economy. This chapter argues that a close analysis of the role of credit and finance in Marx reveals the two sides to be neither formally nor conceptually incompatible. Attention to the complexity of the money form – which as a medium of circulation exists as a generality, internal and other to capitalist forms – illuminates central insights from both Marxism and post-Marxism.

The second part of this chapter, 'Money as Money', draws upon the findings taken from the *Primo Maggio* debate, combined with a close reading of de Brunhoff's significant work *Marx on Money*.[19] In doing so, the chapter mobilises a theory of money as an 'immanent externality': a necessary logical position as the medium of reproduction of capital's abstract form. The logical method of the book is established through reflection on the immanent and external nature of money in its facilitation of capital's abstract movement. Such reflection shows that, value logically, capital requires what is ontologically other to it as a medium of its reproduction. By combining an intellectual history of Marxian debate with sustained theoretical reflection, this chapter develops a genealogical account of de Brunhoff's concept of 'immanent externalities', which, I argue, is essential to understand the ontology of capital.

Chapter 3, 'Fetish Character', establishes a clear interpretation of the fetish character of capital's social form on which the book's full argument relies. This interpretation is necessary to grasp the nature of subjection within capitalist social relations, since it is the fetish character that articulates the specificity of the impersonal abstract social relations that are unique to capital's social form. There is a well-developed existing literature that addresses the fetish character of capital from the point of view of a logical value theoretical reading. Significantly, I. I. Rubin did so with remarkable lucidity in his groundbreaking book *Essays on Marx's Theory of Value*, in which he established the importance of the fetish for interpreting subjection within capital's social relations.[20] Other literature that has constructed readings of capital's fetish con-

---

19  de Brunhoff 2015.
20  Rubin 2016.

sonant with this chapter's argument concerning the dialectic of value forms has come from authors, Elena Lange, Michael Heinrich, Christopher J. Arthur, Stavros Tombazos, and Massimiliano Tomba. While this literature, discussed and applied in the chapter, offers ample material from which to construct a sound interpretation of the fetish character in Marx, the existing scholarship does not systematically reckon with the category of the person. Such a systematic application is hence deployed here as a means to establish a Marxian framework that can address current forms of capitalist domination, where individuals are subject to capital in a variety of ways.

The 'person' is systematically applied here despite Marx's own lack of systematicity regarding the category. This is necessary, I argue, to understand the logical form of relationships internal to capital's reproduction, where non-capitalist social relations are at work in relation to capital's fetishism. It is also necessary to distinguish 'personal' from 'impersonal' domination. So far, however, there has been no clear exposition of the category of the person within Marxian theory. Evgeny B. Pashukanis, in *Law and Marxism*,[21] came close to formulating a clear outline in his observation that commodity fetishism requires a legal fetishism, manifest in the category 'person' as endowed with a will. Étienne Balibar has come closer still. This is revealed in his theory of persons in terms of the 'juridical mask'.[22] However, neither Pashukanis nor Balibar systematically applied the category onto Marx's critique. This chapter seeks to rectify this absence, undertaking an interpretation of capital's social relations as a whole through a systematic use of 'the person'. This is done to better understand the forms of subjection within social reproduction, where the human individual is subject to capital's abstract forms both directly and indirectly, through 'impersonal' and 'interpersonal' forms of domination.

Chapter 4, 'Time and schemas of Reproduction', examines the place of circulation in *Capital* Volume II to discuss how the formal mode of capital circulation – which takes place in expanded reproduction – relies on interruptions of capital's valorisation in order to self-reproduce. Such interruptions are non-capitalist forms that act as a medium of reproduction, taking various forms, from money hoards to material and human life. The reproduction of capitalist social relations engages what is other to capital through the distribution of extraction, excretion, production and consumption in circulation. In doing so, capital's abstractions connect practices that are concretely or temporally and historically disconnected. This chapter shows how the tension

---

21  Pashukanis 1989.
22  See Balibar's chapter 'The Social Contract Among Commodities: Marx and the Subject of Exchange' (Balibar 2017).

between concrete life and capital's abstract form work together in practice within the realm of circulation, in which there is a complex synchronicity and non-synchronicity of temporality between distinct forms and practices. The reproduction schemas in Volume II reflect how the concrete imposes natural limits to the movement of capital's abstract forms.

While existing literature looks at *Capital* Volume II from a value-theoretical inflection, including the collection edited by Christopher J. Arthur and Geert Reuten, *The Circulation of Capital: Essays on Volume Two of Marx's Capital*, which contains essays from Fred Moseley, Martha Campbell, Patrick Murray, Tony Smith, and Paul Mattick Jr as well as essays from the editors (all astute readers of Volume II in their own right),[23] there is scant written analysis that applies the volume to concrete implications of capital's reproduction in circulation. A salient exception is Amy De'Ath's recent entry on reproduction in *The Bloomsbury Companion to Marx*,[24] which explores the current interest in this aspect of Marx's thought in the context of social reproduction theory. Chapter 4 builds on De'Ath's work to further elaborate a robust account of circulation and reproduction.

The chapter proceeds to examine the reproduction of elements that themselves interrupt the reproduction of capital. Thus far, little Marxian analysis has done so. In providing such an examination, the chapter exposes two concepts of reproduction at work: one of the capital's abstract form (engendered in the reproduction of the relations of production), and the other of concrete life (both human and nature). The chapter claims that these respective concepts of reproduction can only be understood with attention to the temporal concepts deployed by Marx in his presentation of circulation in Volume II. The articulation of these two modes of reproduction is then deployed in the last chapter to subsequently articulate Marx's two concepts of 'life'.

The book's argument culminates in Chapter 5, 'Marx's Social Theory of Reproduction', which clarifies the nature of the Hegelian grounding of Marx's idea of the 'life process of capital'. This is analysed to demonstrate that Hegel's concept of reproduction provides a better understanding of the inner tension between capitalist interiority and exteriority in Marx. Of course, existing scholarship has addressed the category of 'life' in Hegel, particularly within *The Science of Logic*.[25] Recently, this was most systematically addressed by Karen Ng in her book *Hegel's Concept of Life: Self-Consciousness, Freedom, Logic*.[26] Convin-

23  Arthur and Reuten 1998.
24  De'Ath 2018.
25  Hegel 2010.
26  Ng 2020.

cing analysis of Hegel's concept of life also appears in Robert B. Pippin's *Hegel's Realm of Shadows: Logic as Metaphysics in "The Science of Logic"*.[27] However, neither of these works consider Marx in detail. In a more Marxian register, Mark E. Meaney has tackled the presence of the Hegelian category of life in the *Grundrisse* in his book *Capital as Organic Unity: The Role of Hegel's* Science of Logic *in Marx's* Grundrisse.[28] Yet, even with Meaney's work, a gap persists in terms of Hegel's 'life' within *Capital*.

In Chapter 5, the retrieval of Hegel's category of life contributes the claim that Marx uses two distinct concepts of life in his critique. One concept is used to explain the abstract life process of capital, the other the concrete life of humans and nature. To develop this thesis, the chapter examines Marx's inheritances: the philosophy of life in German idealism and concepts of life derived from the natural sciences burgeoning at the time of *Capital's* writing. This chapter reworks these legacies to deploy an interpretation of Marx that reconstructs the tension between capitalist and non-capitalist forms, where capital's abstract forms require non-capitalist forms as means of reproduction. As argued in previous chapters, what is other to capital is logically and concretely understood to retain an element in which it reproduces itself for its own sake. This represents an ontological distinction within the logical process of reproduction that permits us to analyse the extent to which the concrete remains independent. Consequently, it is then possible to articulate how this independence can be retained and developed for purposes other than the reproduction of capital's abstract forms.

To summarise, the argument of this book is constructed through three interrelated elements, which the five chapters substantiate from differing perspectives. The first element is that the capitalist mode of production, as a historical formation, depends on a combination of modes of subjection to reproduce. This includes both impersonal relations of capital's fetish character and non-capitalist interpersonal relations. It is argued that the articulation of the proportion of these forms of domination evolve historically in a way that is nonlinear, with the financialisation of capital reflecting a counterintuitive return to the centrality of interpersonal forms of domination.

The second is that the basis of the capitalist reproduction process is constituted by the multiple temporalities of monetary circulation, which involves the circulation of credit money and therefore future constraints on subjects. When capitalist reproduction is understood this way, it becomes apparent that

---

27  Pippin 2018.
28  Meaney 2002.

the site of the central contradiction of capitalism is displaced from the Marxian emphasis on production-wage-labour relations. Although production-wage-labour relations are the conditioning character of capitalism,[29] to assume production as the sole – or even primary – site of capitalist contradiction is a reductive reading of the logic of social relations. The displaced site of contradiction within capitalist social relations ought to be located in the tension between capitalist and non-capitalist elements of capital's circulation. In this book, the non-capitalist[30] elements are termed 'immanent externalities', as they are required for formal circulation to take place and provide the basis for capital's reproduction processes, which are both produced – or shaped – by capital and remain other to capital.

The contradiction between capitalist and non-capitalist elements of capital's circulation symptomatises a general contradiction underpinning Marx's theories of exploitation and alienation. This is the incompatibility between two concepts of life – the life of capital and that of human life and nature – the former being capitalist and the latter non-capitalist. This proposition constitutes the third element of this book. The incompatibility between the life of capital and human life and nature is presupposed by recent theories of social reproduction, which intersect gender, ecological and race-based domination with the logic of capital's abstractions. Consequently, these accounts have made the contradiction between capital's self-reproduction and the possibility of the reproduction of human life and nature glaringly apparent.

The theoretical framework offered through the application of these three elements permits an enriched understanding of the structural interconnection between three prevailing forms of exploitation and social domination: social reproduction, ecological degradation and finance capital. Literature that derives these issues from the structure of capitalism has so far treated the issues relatively separately, lacking a logic of their co-production. There have been few attempts to consider the structural interconnection of the three from the point of view of a critique of political economy. This book constructs an adequate framework from which to understand their co-production. To do this, it reconstructs Marx's project from the perspective of all three volumes of *Capital*, addressed not from the point of view of the 'hidden abode of production' but from what forms the conditions of the possibility of production: reproduction.

---

29   The centrality of living labour, the labour theory of value and the exploitation of labour are all understood to be essential features of capitalism in this book.

30   'Non-capitalist' refers to relations that are not reducible to those of exploitation of unpaid labour and the corresponding movement of value forms that are a product thereof in the sphere of production.

CHAPTER 1

# Fictitious Capital and the Re-emergence of Personal Forms of Domination

## Introduction

At the beginning of Chapter Four of *Capital* Volume I, 'The General Formula for Capital', Marx distinguishes between personal and impersonal relations of domination in a footnote:[1]

> The antagonism between the power of landed property, based on personal relations of domination and servitude, and the power of money, which is impersonal, [*unpersönlichen*] is clearly expressed by the two French proverbs, '*Nulle terre sans seigneur*' and '*L'argent n'a pas de maitre*'. *'No land without its lord' and 'Money has no master'.[2]

Impersonal domination is the form of domination generated by the fetish character of capitalist societies, where social relations objectified in things dominate and act on people. People relate to each other indirectly through the mediation of things, making this form of domination impersonal or determined by a 'rule by abstractions'.[3] Marx articulates the social form behind the rule by abstractions in his theory of valorisation. Therein, Marx demonstrates how value is extracted from labour time to create surplus value: the premise for the accumulation of capital. This process of abstraction endows labour with both an abstract and concrete existence. Abstract labour, the premise of the mode of abstract domination, is labour measured by time and appropriated into value; concrete labour is the actual physical labour that produces things or services. Marx's theory of valorisation relies on the understanding that, in capitalist social relations, human sociality is objectified in and redirected through the circulation of value embedded in commodities, facilitated by the abstraction of labour time into value. Valorisation is a process mobilised by the exploitation of the commodity 'labour power', which is directed towards perpetually increasing the production of capital (or realised value).

---

1 Notably, despite the vast use of 'impersonal' in the secondary literature on Marx, this is the only place in *Capital* where the term appears (Marx 1990, p. 247).
2 Ibid.
3 Marx 1973, p. 164.

Increased financialisation, however, complicates the valorisation process. Interest-bearing capital, the formal motor of financialisation, is money in circulation that has not been valorised through the process of abstraction, where value moves between its forms in production and circulation. Money, in this case, is a form representing un-valorised value. This form of money (un-valorised value) is referred to by Marx as 'fictitious capital'. Fictitious capital is capital that has been credited and represented in the form of money. It represents future-valorised capital, or capital that is not yet valorised. Hence, fictitious capital exists in separation from the valorisation process. While fictitious capital is a form of money endowed with a higher order of abstraction – which therefore represents the mechanism of impersonal forms of domination – this chapter will demonstrate that its increased circulation within capital's reproduction process, counterintuitively, results in the re-emergence of interpersonal forms of domination.

The core premise of this argument is that since there is no realisation of capital without valorisation [Verwertung], there is also no realisation of capital without circulation, where capital is accumulated from production and therefore from labour. Because fictitious capital does not circulate through the production process, there is no moment of realisation of this form of capital. This means that the forms of abstract domination imposed by the circulation of value between its forms do not fully determine the nature of the particular social relationships involved in interest-bearing capital or credit money. This leads to a second proposition: credit necessitates debt and therefore the subjects or bearers of this operation are not merely involved in capitalism as subjects to the value form. They are also creditors and debtors and therefore subject to another form of power relation. Finally, building on these two points, this chapter argues that the power relation involved in credit operations has a personal dimension of dependency, which is, in turn, premised on the realisation of capital through the valorisation process. Furthermore, this personal form of domination, residing at the core of the debt contract, is not wholly without its own abstractions, for 'to receive a "juridical qualification" [the contract] is inscribed in a legal system marked by its abstract universality'.[4] Thus, abstraction qua abstraction is not what gives the value form its impersonal character. Impersonal domination is the result of the absence of personalised agents representing the abstraction of the value form.[5] In contrast, legal contracts, are represented by owners (or groups of owners) who

---

4  Bhandar and Toscano 2015, p. 11.
5  The capitalist is 'capital' personified and not a personalised agent for capital.

are persons.[6] Hence there are different modalities of abstraction at work in Marx's account of capital,[7] and what makes the form of abstract domination imposed by the value form impersonal is not merely abstraction itself. The abstraction becomes impersonal when the appearance of an actor representing the abstraction is not a person but a thing. In capitalist relations, this 'thing' is the form of appearance of value, a product of human labour, which, moving from one commodity form to another, becomes an independent actor or what Marx terms the 'automatic subject'.[8] In this book, the 'impersonal' is therefore construed as distinct from – but not incompatible with – more common uses of the term, as can be found in writings by Moishe Postone, Michael Heinrich and Søren Mau,[9] where impersonal relations of employment, contracts and competition are opposed to personal relations of servile obedience and feudal obligation. The specificity of the 'impersonal' differs in this interpretation in so far as the 'impersonal' denotes the specifically capitalist mode of exploitation without assuming the interpersonal aspects of that mode to be internal to that form of exploitation. The 'impersonal' is here interpreted as a distinct formation that exists alongside other modes of domination that are interpersonal, permitting the development of relations within capitalism such as contracts and competition. 'Interpersonal' modes of domination, for their part, do not lack an abstract character: they are relations that are mediated between legal persons without the direct arbitration of the value form (characterising impersonal capitalist relations). Both impersonal and interpersonal forms of domination are necessarily present within capitalist social relations.

The persistence of interpersonal forms of domination is not only based on the relation of the debt contract but also on the personal relations that uphold the ability for society to reproduce the labour power that will account for the repayment of debt in the future. The personal forms of domination implicated in social reproduction are placed under increased pressure by the heightened circulation of fictitious capital. This is because society, as a whole, is struggling not only to produce capital in the present but also to make up for the fictitious capital that is in use. As such, capitalist society has come to presuppose the reproduction of labour power and the conditions of production in the future *as well as* the present. However, these personal forms of power

---

6 What I mean by 'personal' and persons in relation to the 'impersonal' is developed in Chapter 3, 'Fetish Character'.
7 Bhandar and Toscano 2015, p. 11.
8 Marx 1990, p. 255.
9 See Heinrich 1993; also see Mau 2021, pp. 3–32.

relations come into being by the very process of exchange that is described abstractly by Marx as completely impersonal and formal based on the 'high-level *logic* of abstraction'[10] intrinsic to the value form. This is because the social relations of credit operations in capitalist societies are built on the social relations of the value form. Hence, personal forms of dependency brought about by fictitious capital's exclusion from valorisation do not index an absence of abstract forms of domination, nor do they imply that the juridical qualification behind the debt contract is not itself a form of abstraction. Rather, the increase of personal forms of dependency means that the abstract forms are upheld by personalised relations of domination functioning in the realm of the reproduction of capitalist social relations. These personalised relations include the reproduction of labour power (in the everyday, personal relationships that sustain one's ability to labour which are often distributed through forms of care). In this way, the conditions that permit the re-emergence of the exploitation of forms of domination, which have been manipulated to uphold reproductive aspects of capital production, reside within the process of capital's valorisation itself. With 'fictitious capital', directly personal power relations come to the fore, facilitated by the impersonal domination of commodity fetishism. As David Harvey notes, 'the credit system becomes the locus of intense factional struggles and personal power plays within'.[11] Shulamith Firestone, for her part, rightly predicted that with the increasing technologisation of the mode of production that facilitates the unprecedented fictitious capital in circulation, 'cybernation may aggravate the frustration that women already feel in their roles'.[12] With the dominance of 'fictitious capital', social reproduction increasingly relies on personal relations, while power structures ordinarily considered external to the production process re-emerge within capital's valorisation in a renewed way (yet under different structural relations).

To show how fictitious capital re-centres these forms of 'personal relations', it is instructive to first understand what Marx meant by fictitious capital and how fictitious capital relates to the impersonal domination exemplified by money. This chapter will therefore first, in section *1.1 Fictitious Capital*, establish Marx's use of the concept fictitious capital in its representation of a particular appearance of the money form sustained by future value. Following this, in section *1.2 Fictitious Capital and Value Form*, the chapter examines the role fictitious capital plays in the context of Marx's theory of capital valorisation by undertaking

---

10  Bhandar and Toscano 2015, p. 11.
11  Harvey 1982, p. 287.
12  Firestone 2015, p. 202.

an analysis of value forms. This section locates the place fictitious capital takes within the movement of value as a mediator between impersonal domination and interpersonal forms of domination.

The interpretation of Marx undertaken here will subsequently argue that the reason for fictitious capital's mediation of interpersonal subjection with impersonal abstract form arises through a distinct temporality that requires future value – and not past labour – to sustain its presence. The presence of future value is then shown to be secured through a juridical contract between persons. Finally, in section *1.3 Social Reproduction and Personal Domination*, the chapter will synthesise its findings regarding the circulation of fictitious capital with an analysis of social reproduction. This final section will add the logical extension of the relation between social reproduction and fictitious capital to Marx's critique. The latter is understood as a form that subjects its bearer to the money form, beyond the practice of exchange, in a relationship of dependency that is interpersonal. By rendering its bearers responsible for future value, fictitious capital is here argued to secure not only the domination of the person but the entire network of social reproduction. This chapter argues that interpersonal domination is a condition of possibility for the reproduction of the future labour held in anticipation by the circulation of fictitious capital.

## 1.1    Fictitious Capital

The concept of fictitious capital, developed in *Capital* Volume III, has largely been left idle and loosely defined. Recently, however, it has become the focal point of debates concerning 'financialisation'.[13] What Marx means by fictitious

---

[13]  The literature I am primarily referring to occurs in discussions and debates surrounding the works of Costas Lapavitsas, David Harvey, David Graeber, François Chesnais, Maurizio Lazzarato and Thomas Piketty, to name a few thinkers contributing overlapping and competing perspectives. 'Financialisation' is chiefly understood here as a shift in the capitalist economy, whereby the expansion of the circulation of interest-bearing capital has acquired increased weight, manifest in financial forms of profit and revenue. Financial operations extract profit without producing or through the extraction of value produced in another context within the economy: finance is not understood to produce value but as a means of extracting value, either as surplus through rents or in the creation of fictitious capital. The term – although indicating a qualitative shift in the way that capitalism extracts profit – is often used as a periodising category to denote a tendency within the development of capital relations. The periodising function of financialisation in this book is generally in line with Giovanni Arrighi, who sees financialisation as a reoccurring and uneven tendency intrinsic to the development of capitalist

capital is the practical use of one form of value – as it appears in the money form – more than once. The multiple use of a single thing involves a 'fictional' element. This dynamic occurs in the case of credit operations, where credited value is at once owned by the creditor and practically used by the debtor (who could be an individual or a manager of money representing a banking or investment operation). Fictitious capital is lent money that represents a title to future value, functioning as a fictitious appearance of the original money still owned by the lender. The copy mimics the original with the exception that its valorisation is suspended. The suspension of valorisation is constitutive of fictitious capital because its valorisation (repayment) is the cause of its disappearance. By charging interest, this form of money can generate more money and is described by Marx as $M-M^1$, which is opposed to $M-C-M^1$, where more money is generated due the mediation of commodity production.

The twenty-first century has seen a cyclical moment in the capitalist mode of production, whereby the intensification of the function of the credit system and the use of 'fictitious capital' has been accompanied by technological change. This intensification has changed many of the ways in which capital generates profit through the exploitation and subjection of members of capitalist society. This change in quality of capital exploitation is determined by a shift in production's temporality: production is no longer merely that which makes value possible, as an aspect of value's prehistory, but is assumed to exist in the future to account for rising proportions of fictitious value.

When a single form of value exists twice, the doubling of that value entails a temporal displacement in the process of valorisation. For value to exist twice and to thus contain a fictional element, the second use of the value requires the anticipation of future production; this is understood to represent the 'second' manifestation of value. However, this second value form is not yet valorised,[14] which is why it needs to be represented as a fictitious token or entitlement. If a loan is repaid to the lender, the fictitious aspect of said value disappears. In this regard, fictitious capital is never valorised since it is a placeholder waiting for the appearance of a second form of value to arrive from the future. Fictitious capital first plays a role in Volume I of *Capital*, before Marx develops the role of

---

relations – a tendency logically internal to capital and not a new stage of history. For Arrighi, every productive expansion is followed by a financial expansion. See Arrighi, 1994.

[14] Although exchange has occurred, production has not, and value is valorised through the interrelation between production, circulation and exchange.

the credit system. Credit money first appears in the dynamic of simple circulation as a symptom of the temporality of the development of value as it moves from one form to another in its ultimate production of surplus-value and later capital.

Within simple circulation, there are two central concepts with distinct temporalities: labour time (both concrete and abstract) and the time of circulation. Labour time is operatively used as a measure of value, while the time of circulation is concerned with empirical time or the time it takes for capital to circulate throughout social relations. These two temporalities do not always substantiate the other in a linear way. The non-linearity of the C-M-C relation requires the modification of credit money: 'credit [is] a modification of the commodities/money/commodities exchange, which in a market circulation mode necessarily takes place in a simultaneous manner',[15] i.e. in the case that the commodity buyer does not and cannot pay in hard cash. Due to the expected future cash that the exchange promises, credit steps in not only to play the role of future money but to create the very conditions of possibility for future money. As such, in order to fulfil its role in facilitating commodity exchange, credit money must be non-synchronous[16] with the accumulation of capital. By necessity, credit exists within an aleatory temporality to compensate for temporal contradictions in the logic of capital accumulation. As de Brunhoff has shown, credit does not share temporality with capital.[17] Thus, within the process of simple circulation, credit introduces time as a formal determination of value in its money form. The temporal discrepancy between fictitious capital and valorised value (both appearing in the money form) constitutes a disjuncture that comes into being to compensate for temporal inconsistencies found in the social form of capital accumulation.

The function of credit money within the context of simple circulation is referred to as 'commercial credit' and forms the basis of the development of the credit system.[18] As a direct product of the use of money as a means of payment, commercial credit (or simple credit money) no longer corresponds to the simple money form and is not yet representative of the role of credit within the credit system. Hence, de Brunhoff's correct assessment that commercial credit can be found,

---

15    de Brunhoff 2005, p. 211.
16    Here, to be non-synchronic is to exist as a non-capitalist temporality. The relationship between temporality and non-capitalist forms and practices will be developed throughout the book.
17    See de Brunhoff 2015.
18    Marx 1991, p. 400.

> ... on the borderline between the monetary system and the credit system. Incorporated into the latter, it introduces into it the contradiction inherent in the function of money as a means of payment, which represents simultaneously the ultimate dematerialization of money and its re-embodiment.[19]

When credit money plays the role of money as means of payment, credit money clearly reflects its role as a form determination (and not an objective thing with inherent value). What is necessary about the money form operatively is that something stands in for it to ensure its role. However, money's 're-embodiment' in the form of credit presents a problem for what constitutes its monetary character, for it represents both less and more than money: it represents less in the sense that it stands in for future money, and more in the sense that it is upheld by an additional juridical contract based on the debtor-creditor relationship. The debtor-creditor contractual relationship supplements the basic formal contract that money already abstractly represents.[20] Furthermore, credit money exists through the circulation of debts and is thus the 'non-circulation of money'.[21] Yet credit money has by necessity 'monetary characteristics' as a medium of circulation and a measure of value, albeit a measure of value with a different temporal relationship to production. This temporal difference brings about money's non-equivalence to credit money. That is, the measure of value by credit money cannot be fulfilled directly, because future value (the kind of value bared by credit money) has not yet come into being. Although credit money stands in for future value in a nominal way, it cannot confirm the existence of future value. Hence, while credit money bears the form of value, value is not directly embedded in this form. Accordingly, credit money's monetary characteristics carry a purely abstract significance and do not indicate practical convertibility of value forms. De Brunhoff shows,

> ... the first function of money, that of the measure of values, cannot be directly fulfilled by credit money. In that sense, "credit money" is only 'money in so far as it absolutely takes the place of actual money to the amount of its nominal value.' But this convertibility has only

---

19   de Brunhoff 2015, p. 81.
20   The category 'contract' is used in two different ways in this sentence. The first refers to a 'juridical contract' and the second is a symbolic 'social contract'.
21   de Brunhoff 2015, p. 83.

theoretical significance; in normal times it does not in any way imply an effective convertibility.[22]

Thus, while it is clear that credit money can only act as money form due to its monetary characteristics, credit money cannot merely replace the function of money. This is because the temporal inadequacies will ultimately reveal themselves in a time of crisis. When there is no longer speculative evidence for the reproduction of future value, credit money loses its 'credibility' as a stand in for money form and will need to be replaced by the money form proper. When there is a lack of speculative anticipation that the future will bring the reproduction of value, credit money is recovered *en masse* leading to disastrous effects in the lives of debtors, such as housing foreclosures.

Fictitious capital, confirmed by a juridical relationship between two parties, essentially splits the function of the money form in two. This split is governed by a legal contract acting as a claim over money. Here, there are two opposed legal subjects, the debtor on one side and the creditor on the other. Money takes on two forms. The first form, in the case of the creditor's relationship to the money, is one in which the money becomes a 'unit of account' for a sum that is due in the future. In the second form, the debtor holds money that represents their future labour. It is this future labour that will then produce money to act as a means of payment for the settlement of the debt. De Brunhoff shows that this is 'a specifically constrained *institutional* relationship that is governed by contract and law',[23] added to the always already institutional nature of money, legitimised by the state in the first instance. The institutional relationship governed by contract and law is necessary for the temporality of the credit relationship:[24] The contract functions to legally uphold the manipulation of future time to insert money into the circulation process. Credit money, as an appearance of the money form, would have been the result of the process as valorised value, but it ultimately is not. Credit money is therefore neither present in earlier stages nor persistent after the debt's repayment.

To act as a money form, credit money sustains its presence by securing a co-dependent relationship of subjection between the creditor and the debtor, or between the owner of money and the owner of future money. This co-dependency although might seem to be secured by the dynamics of capital

---

22   de Brunhoff 2015, p. 84.
23   de Brunhoff 2005, p. 211.
24   de Brunhoff 2005, p. 212.

accumulation premised on the value-form, is in fact formed by a legal contract. Here, credit money as a non-contemporaneous form of value's appearance becomes implicated, as the bearer of the form,[25] to uphold it (despite its temporal lag). This brings full circle the two concepts of time developed so far – the abstraction of labour time as measure and the empirical time of social practice in the realm of capital's circulation – through their shared structural role in implicating the subject in the abstraction of the value-form. 'Labour time' supplies the content of the measure of value in its form of representation as money, while the debtor supplies the solution to circulation's temporal inconsistencies by supplying future labour time as content to the measure of its value as credit money. Thus, there is no function of 'time' as an abstract form underpinning capital accumulation without subjection.[26] However, there is more than one kind of subjection at work, corresponding to different temporal categories: subjection to forms of value on the one hand and subjection to one's future valorisation on the other. The latter form anticipates future subjection to the value-form based on the 'personal' memory of one's past. This indicates a suspension of subjection to the subsumptive function of value-forms – a function that, while latent, is replaced by a different kind of subjection: subjection that is based on a personal relationship of dependency comprising one's anticipation of the future.

Subjection to the debt contract retains the legal person internal to commodity fetishism (the person of the exchange relationship, or a property owner) without the mediation of the commodity form, making it an interpersonal relation rather than an impersonal one.[27] Significantly, the temporality of money and its subsequent forms of appearance provide a vital link to the abstract self-movement of the value-form (based on abstract labour time), forming the temporality of capital and the temporality of the individual subject, who is rendered a person by a legal contract. The individual subject is implicated in the temporality of capital through the time of social practice that, in capitalist social relations, is structured by the temporality of the movement of value. However, practices undertaken by a person to reproduce their lives can persist without the movement of the value-form (the capital relation), under a

---

25  The bearer is subject to the capital exchange abstraction.
26  Temporality implicates the subject by creating the conditions for the measurement of labour, thus facilitating the alienation of the subject by enabling the abstraction of value form that structures social form.
27  The distinction between a personal and impersonal relation is defined in detail in Chapter 2, 'Fetish Character' in the section 'Personal and Impersonal Forms of Domination'.

different, non-capitalist set of social relations. Therefore, the temporal site of personal forms of domination can be best understood from the point of view of the individual subject, as a legal person, and not from the perspective of the impersonal abstract domination of value.

## 1.2   Fictitious Capital and Value Form

In *Capital*, there are different expressions of the commodity form, including the object commodity of production, labour as commodity and the money commodity. Therefore, what makes a commodity a commodity is not its materiality but its social form. However, different appearances of the commodity form will constitute different social relationships, with different temporal structures, which occasion different outcomes. For example, in the case of exchange, the only difference between a material product and an immaterial one, such as a service, is that they have a different relationship to the timing of production and consumption. As Michael Heinrich accurately describes, 'the difference between services and physical objects consists of a distinction of the *material* content; the question as to whether they are commodities pertains to their *social form*, and that depends upon whether objects and services are exchanged'.[28] The presence of so-called immaterial labour does not oblige a re-examination of Marx's value theory, since the social form remains the same. With immaterial labour, we see a temporal shift because production and consumption happen simultaneously rather than through the mediation of an objective (and thus future) use value. Meanwhile, there remains the mediation of use value with the phenomenon of instantaneous consumption. In this dynamic, the commodity does not require its objective form. This is also true of the workings of the money commodity that does not require an objective form (a gold standard, for instance) to operate as a commodity fulfilling the role as a universal equivalent.

Money has a use-value (it is used as a mode of value's circulation) and an exchange-value (it represents a quantity of value), and therefore expresses the commodity form. However, when functioning as fictitious capital, money's use-value becomes its potential to create greater value. Hence, its value is based on a self-fulfilling irrationality. As Marx describes, 'the value of its value is that it produces greater value'.[29] Fictitious capital has a use-value but no definable

---

28   Heinrich 2012, p. 44.
29   Marx 1991, p. 354.

value, indicating that fictitious capital does not represent the material production process where the embodiment of labour is embedded in the form, such as a commodity.[30] Harvey elaborates that the primary role of the money commodity is to function as 'a medium of circulation as its use-value is that it facilitates the circulation of commodities'.[31] Therefore, as Harvey continues, 'from the standpoint of a pure medium of circulation, money can equally well take any number of forms'.[32] It is because money represents the universal equivalent of exchange value that it is essentially opposed to other commodities: 'money assumes an independent and external power in relation to exchange because, as the universal equivalent, it is the very incarnation of social power'.[33] Thus, it is evident that the reliance on gold as a standard for the objective form of money was not a necessary component of Marx's critique. Rather, the gold standard was a formation assumed in practice as a stage in the historical development of money as value form.

Using this same logical framework, Heinrich has observed that it is not a problem for Marx's theory of value to replace the objective use-value component of a commodity with an exchange of property, due to a shift between the time of production and consumption. This observation is significant for conceiving the relationship between value form and labour even in the case of the exchange of fictitious capital, where Marx shows that capital accumulation 'appears' to occur without the mediation of production. This is because the M-M$^1$ formation nonetheless relies on labour to generate value in the first place. However, it might be asked, what happens when value ceases to be predominantly extracted from the site of capital's accumulation process? By neglecting the repercussions of a shift in the respective timing of production and consumption, Heinrich does not examine the possibility that the shift of past labour to the future has potential social implications that exceed what can be accounted for through value forms. A shift in the timing between production and consumption differentiates the object commodity of production (as something containing abstract labour and therefore value) from the money commodity of fictitious capital. In fictitious capital, the producer receives a sum of equivalence for an exchange value before they produce what will account for this exchange value, while the creditor essentially owns the speculative future production. This formulation, marked by a change in timing, means that the relationship is no longer a relationship of exchange where

---

30   Harvey 1982, p. 259.
31   Harvey 1982, p. 354.
32   Harvey 1982, p. 244.
33   Harvey 1982, p. 245.

equality is constructed between two things, as Macherey puts it, 'so as to have value appear and destroy things'.[34] This is because the subsumptive force of exchange in the construction of equivalents, in the movement towards valorisation, is only retroactively constituted (or made real) after capital has been valorised in the final exchange on the commodity market. While fictitious capital acts as exchange value when used on the market, and creates profit through interest, does not create value. This is due to a difference in timing: When fictitious capital is put to use, its valorisation has not yet occurred. While money engages in relationships of exchange, the bearer of fictitious capital remains tied to the money form beyond the performance of this exchange and in a relationship of dependency that is directly personal (or interpersonal) because it involves social relations that do not need mediation by the value form to uphold legitimacy. This is symptomatic of the latent effectivity of value before the final exchange of the commodity.[35]

A temporal shift in capital's abstract relation entails the restructuring of the value form's relationship to its empirical social referents. This involves a refiguration of the nature of subjection. For fictitious capital to exist, the moment of valorisation of capital must be withheld. With valorisation withheld, valorised value is replaced by a person who represents the guarantee that – in the future – labour will be undertaken. This bearer of fictitious capital represents, as Marx puts it, a 'claim' or 'legal title to future production'.[36] If we consume first and produce later, the latter production is determined by a form of subjection that ensures future labour. With this formal shift, a personal relationship of dependency is upheld where domination between persons is extended over time and formalised in a contract. As in the case of landed property, the owner of the fictitious capital is 'Lord' up until fictitious capital becomes valorised. Once fictitious capital is valorised, the directly personal contract disappears and persons once more enter the world of impersonal exchange where persons relate indirectly through the mediation of things.

Fictitious capital's evasion of valorisation, and thus capital realisation, accords with the ontological claim that Marx's theory of valorisation makes: there is no value without the expenditure of human effort through labour (which is then represented abstractly by the measurement of labour time used to produce value as an abstract form). 'Surplus value' appears as the portion of the

---

[34] Althusser et al. 2016, p. 199.
[35] The belief that fictitious capital will in the future produce value means that value form's tendency towards realisation is latently present in the fictitious capital by imposing future actualisation of the valorisation process that is not actual in the present.
[36] Marx 1991, p. 599.

working day that seems to be paid. This is due to the wage's obfuscation of unpaid labour time. Capital appears to accumulate from nothing because the extraction of value from labour exceeds the amount paid in wages. Fictitious capital is fictitious because its creation occurs independently from the production process, and it therefore does not have a premise in labour. In this way, fictitious capital has come to 'evade the conditions of the circulation of capital'[37] and thus commodity circulation and production.

The increased use of fictitious capital throughout the twentieth century,[38] eluding the conditions of production, seemingly contrasts with the intensification of industrial factory work and the proliferation of labour time functioning as a measure of value. This has been driven by investments in technology that have limited the need for labour. This production of relative surplus value over time leads to a depletion of the primary source of value, which is labour producing absolute surplus value. However, on a broader scale, capitalists that cannot invest in technology are forced to compensate by devaluing the cost of labour power to compete with the change in socially necessary labour time that has been redefined by relative surplus value. This is a driving factor behind the falling rate of profit that becomes compensated for by finance capital through the evasion of production and the extraction of profit in circulation. Postone astutely characterised this dynamic as follows:

> With the increase of productivity, you have increases in material wealth greater than increases of surplus value but surplus value remains central to the system. This means that the system generates accelerating production, using the accelerating use of raw materials for smaller and smaller increases in surplus value.[39]

Massimiliano Tomba has further pointed out that the claim that there has been a decrease in labour time that facilities the abstract labour embedded in the value form is one sided. Tomba rightfully shows that this can be said to be true 'only in relation to the productive force and intensity of a socially necessary labour-time [where the] average does not pass through the north-western

---

37  de Brunhoff 2015, p. 94.
38  This general trend of the increased circulation of capital is well documented by Costas Lapavitsas in his book *Profiting Without Production*, published in 2013; François Chesnais's *Finance Capital Today: Corporations and Banks in the Lasting Global Slump*, published in 2016; Michael Roberts's *The Long Depression: Marxism and the Global Crisis of Capitalism*, published in 2016; and, with specific focus on the terminology of fictitious capital, Cédric Durand's *Fictitious Capital: How Finance Is Appropriating Our Future*, published in 2017.
39  Postone 2016.

and non-US axes'.[40] This dynamic lies at the centre of Dipesh Chakrabarty's argument in *Provincializing Europe*.[41] There, Chakrabarty describes how, with the majority of socially necessary labour time occurring in the Global South, the dominant temporality of global capitalism needs to be understood from a renewed perspective. Labour time internal to the generation of value in capitalist societies has in fact increased, yet it has been greatly devalued by low wages in the Global South. Simultaneously, in countries where capital accumulation is centrally organised, there has been large scale deindustrialisation – largely replaced with unemployment and low-paying service industry jobs – while there has been increased creation of value through the $M-M^1$ relation of fictitious capital or financialisation. As Postone has noted, in this dynamic, the increase of productivity occurs without an increase or even decline of surplus value, which nonetheless remains structurally central. Here, Postone argues, it is evident that 'value's growing inadequacies as a measure of social wealth remains the necessary structural presupposition of capitalist society, and this is the basis of capitalism's fundamental contradiction'.[42] This contradiction is between value as the product of labour and the potential for other forms of wealth to develop, such as fictitious capital. This represents a historical contradiction whereby capitalist social relations create the conditions of possibility for generating new forms of wealth, while the general realisation of this wealth is simultaneously constrained by old forms (i.e. value's reliance on labour).

Fictitious capital is speculative in nature, as it exists as the placeholder for a dialectical result of the reversed movement between the universal form of exchange (money) and labour. To reiterate, this process is premised on fictitious capital's disappearance at the moment of valorisation. When fictitious capital is repaid, instead of valorising capital the repayment eliminates the fictional element and the fictitious capital no longer exists. Hence, fictitious capital does not exist beyond the realisation of the speculative proposition that future labour will supply its repayment. Thus, fictitious capital is always money advanced for future labour not yet secured: it will never be realised.

Drawing on Hegelian terminology, fictitious capital represents a determinative 'nothingness' that negates the contradiction that contains within it a weakness that will fall apart by the end of its movement. The function of fictitious capital is at once less and more than the actual function of money as commodity, because 'fictitious capital' does not contain the same boundaries as other money commodities nor does it require the same labour relations. However,

---

40  Tomba 2013, p. 168.
41  Chakrabarty 2007.
42  Postone 2007.

fictitious capital's speculative nature is a derivative expression of another type of exploitation: production and the corresponding extraction of surplus value. The doubling of fictitious capital is an appearance of a sum initially extracted as surplus value from wage labour. When the expansion of commodity values is unable to maintain pace with the preceding creation of fictitious capital (in overaccumulation or fictitious accumulation), we find a demand for liquidity that causes inflation. When inflation cyclically occurs – which is the appearance of the structural contradiction between capital in its monetary form and capital in its commodity form – the illusion of fictitious capital no longer suffices as a placeholder for valorised capital.[43] Marx considers this to be major cause for cyclical crisis in the history of capitalism: the crisis being the result of the contradiction. If the use of fictitious capital outweighs the conditions of production, the mystification will concretely reveal its material inadequacies. Yet, the more strain that is placed on the production process (in the decrease of wages, the outsourcing of labour etc.), the less it is possible for fictitious capital to be repaid, since when production is weak it becomes increasingly difficult for interest on loans, bonds or investments to be recovered, let alone repaid. Furthermore, this suspension of valorisation is the programme of the financier: so long as interest payments are recovered, a permanent suspension of repayment is in their favour.

Meanwhile, the suspension of valorisation engendering fictitious capital imposes a different kind of exploitation of the subject, or bearer of that form of capital, than in wage labour where value is valorised. The repayment of fictitious capital becomes grounded not only in the worker's discipline to spend time labouring but also in their ability to labour in the future. 'Free time'[44] is therefore overdetermined by the struggle to further reproduce labour power. A crucial part of the legitimation of fictitious capital relies on the constitution of a corresponding subject who lives a life pathologically driven towards the ability to repay, and pay for, what has been credited to them. This personal form of domination, unlike capital's impersonal relations, relies on domination to occur outside of the confines of the working day. However, this is not unique to the function of fictitious capital: exploitation of one's whole life is also a central aspect of consumer culture, colonisation, the disciplining of the unemployed, women's unpaid work, as well as explicitly political forms of domination. Thus, capitalism evidently grounds itself in the het-

---

43  For a detailed discussion of overaccumulation, fictitious capital and crisis, see David Harvey's *The Limits to Capital*, particularly in the section 'Finance Capital and its Contradictions' (Harvey 1982, p. 296).

44  Reference to 'free time' is given by Marx in the 'Notebook VII – The Chapter on Capital' (See Marx 1973).

erogeneity of exploitation, whether personal or impersonal. The exploitation of wage labour and one's life in society (which is other to the capital relation) together constitute the contradictory unity of production and its realisation.

### 1.3   Social Reproduction and Personal Domination

Marxian value-form theorisations of impersonal domination have typically overlooked how the premise for the abstraction – the labour commodity – is produced and reproduced. The field devoted to sustained study of labour's production is referred to as 'social reproduction theory'. Social reproduction theory has been developed in contradistinction to socialist feminist theory, which has largely sought to understand the relationship between women's oppression and capital in terms of dual systems of oppression, later revised as intersectional systems to include race, class and gender. This form of analysis implies that each system has its own history, informing aspects of what feminist theory generally terms 'intersectionality'.[45] In the context of feminist thought, Marxism has often been understood reductively as a tool to understand class alone and is not seen as a theoretical field that grasps forms of social exclusion or power structures, such as domination by way of multiple levels of abstraction, or how capitalist social relations depend on non-capitalist forms of domination. In contrast, social reproduction theory assumes that Marxian thought can be extended to encompass not only gender and race but a wide realm of diverse forms of exclusion and domination that determine how one relates to the production process, a process that is understood to impose forms of impersonal abstract domination that cannot be reduced to a class analysis.

The framework deployed here is developed here with a focus on the reproduction of labour power in capitalist societies, based on the observation that while capitalist accumulation relies on labour, it does not produce labour. Furthermore, as Susan Ferguson observes, 'there is no mechanism in the direct

---

45    This term was coined by Kimberle Crenshaw in a 1989 paper, 'Demarginalizing the Intersection of Race and Sex: A Black Feminist Critique of Anti-discrimination Doctrine, Feminist Theory and Antiracist Politics', to explain that black women's experience cannot be understood if the two identities 'black' and 'women' are thought separately. Subsequently, the concept became more widely used after the publication of Patricia Hill Collins's book *Black Feminist Thought: Knowledge, Consciousness, and the Politics of Empowerment*, where she refers to the 'matrix of domination' as interlocking modes of oppression to include other differences such as age, sexual orientation, and class (See Collins 2000; Crenshaw 1989).

labour/capital relation to ensure labour's daily and generational renewal'.[46] As a result, the daily and generational renewal of labour is ensured at the level of kinship structures and individual consumption. These are social relations that are internally structured by personal forms of domination (of juridical persons).[47] The realm of social reproduction regulates daily survival and the reproduction of labour in practices such as eating, sleeping, sexuality, care and child rearing. Lise Vogel explains,

> The bearers of labour-power are, however, mortal. Those who work suffer wear and tear. Some are too young to participate in the labour-process, others too old. Eventually, every individual dies. Some process that meets the ongoing personal needs of the bearers of labour-power as human individuals is therefore a condition of social reproduction, as is some process that replaces workers who have died or withdrawn from the active work force. These processes of maintenance and replacement are often imprecisely, if usefully, conflated under the term reproduction of labour-power.[48]

It is within the general framework of what Vogel refers to as a 'unitary theory of social reproduction' that the reproduction of labour power is understood as internal to capitalist totality. This means that personal forms of domination cannot be adequately understood as pre-capitalist, or post-capitalist: they must be grasped as contradictory forms internal to capitalist social relations themselves. The theoretical premise of a 'unitary theory of social reproduction' is rooted in inherent internal differentiation within the concept of totality. To quote the famous rationale extracted by Vogel, Marx claims,

> The maintenance and reproduction of the working class is and must ever be a necessary condition to the reproduction of Capital. But the capitalist may safely leave its fulfilment to the labourer's instincts of self-preservation and of propagation.[49]

---

46   Ferguson 2014, p. 165.
47   A full definition of this form of domination is given in Chapter 3, 'Fetish Character', in the section 'Personal and Impersonal Forms of Domination'. There, it is argued that personal forms of domination are representative of the 'juridical mask', or the character mask of the person as a juridical by-product of the exchange relation, that, in the dominance of exchange in capital's social relations, is imposed on all individuals as a social form, even if they are not engaged in an exchange.
48   Vogel 2014, p. 144.-
49   Marx 1990, p. 718.

Within Marxian social ontology, what Marx refers to as instincts will, of course, remain symptomatic of historically specific social relations. Hence, it is within the concept of so-called 'instincts of self-preservation and of propagation' that Marx permits his framework of capitalist totality to be augmented with a theory of social reproduction, which concerns those whose lives are determined by personal forms of domination, or who might be excluded from labouring but largely devoted to maintaining labour or the future possibilities thereof.

Social reproduction theory does not base its analysis on the idea that social reproduction necessarily entails a gendered division of labour, Euro-American kinship relations or the organisational foundation of the nuclear family. Rather, it sees capitalism as having also found other ways to ensure the daily maintenance and the regeneration of labour power. This is a process that is maintained 'through hierarchically and oppressively structured institutions and practices, such as private households, welfare states, slavery and global labour markets'.[50] Gendered forms of violence, colonial, racial and the exclusion of anyone who is unable to themselves labour are therefore central to the constitution of labour itself. This is because the site of social reproduction is at once the site that produces and reproduces labour power, and also the site that produces and reproduces life qua life, regardless of whether one is able to labour. Accordingly, any form of exclusion from the marketplace – from racial exclusion and old age to increased unemployment or under-employment – imposes additional stress on the realm of social reproduction. How, then, does fictitious capital's suspension of capital's realisation intensify these personal forms of domination in the realm of social reproduction?

The absence of labour time reveals and brings to the fore the aggregate of personal forms of domination found in the realm of social reproduction, which are always already present but often hidden in production. This is because the subjective domination of the one who bears the fictitious copy of money capital (or credit money) relies on the subjectivity of the bearer of fictitious capital to mediate between production and consumption (this subject has to prove themselves as a one who can produce future valorisation and therefore as a subject who can reproduce their labour power). Generally foreclosed from the possibility of engaging in the actual production necessary for full repayment of debt, this subject is required to stage their appearance as an active, adequate reproducer of labour power. The actual production of labour power is then a state that is upheld in suspension along with capital's valorisation. Here, the

---

50  Ferguson 2014, p. 165.

personal forms of social domination intrinsic to social reproduction become a central justification in the legitimation of capital's production of more value through the M-M$^1$ relation. One must prove that they are able to command the kinds of domination needed for the production of their labour power, or the production of future social wealth, that will stand in for fictitious capital in the future.

'Debtors' are not necessarily individuals but can be nations, corporations or financial institutions, such as banks and hedge funds, who speculate on future production. Thus, power dynamics and their implications vary significantly. However, when much of the money being used in the economy is fictitious, money in use is not generated by production, and thus, on a societal level, the labour needed to uphold it has been transposed from the past to the future. In this regard, the production process upholding this form of capital has been transposed to the future as well. Societally, many of the people who should have been subject to the past process of production are subject to the future process, meaning that they must appear to be able to labour in the future and able to reproduce their labour in the present. If future labour is indefinitely absent, the necessity to reproduce pending labour remains. Hence, whether or not fictitious capital is repaid (much is never repaid but re-bundled and sold as a financial asset), society must nonetheless act 'as if' it will be repaid in the future. Social life must reproduce in the present in such a way as to be able to both continue to exist in the present and to produce individuals who are able to repay debts in the future. Therefore, individuals are forced to invest in their own skills and abilities (their 'human capital'). This can surface in the form of self-discipline (in Foucault's sense).[51] Yet this also has societal effects bearing on the structures of domination that uphold the possibility of reproducing one's labour power, culminating in increased violence toward those who themselves are already excluded from the production process, whose lives are appropriated for the upkeep of the possibility of labour power. This is especially acute in forms of exclusion from labour markets. Here, the personal nature of domination, as the manifestation of the latent abstraction indicative of the value form, infiltrates one's whole life.

'Fictitious capital', as a form that facilitates the renewed emergence of directly personal domination within the process of capital's valorisation, reflects a significant contradiction: Marx claimed that fictitious capital is capital's most fetishised form and therefore its most impersonal. For this analysis, the primary issue is that of whether these combinations of the personal and the imper-

---

51   See Foucault et al. 1991.

sonal displace forms of subjection, rendering them open to struggles against the impersonal domination of capital's abstractions, or whether they produce unprecedented means of legitimising capitalist social relations.

This chapter has presented a proposition which subsequent chapters will confront. The implication of its argument is that interpersonal forms of domination are an immanent product of capital's self-reproduction. Such forms of domination, furthermore, undergo greater stress as the proportion of fictitious capital rises. This occurs when the accumulation of fictitious capital cannot be fully accounted for by the future reproduction of the production process.[52] Capital's abstract forms are consequently able to reproduce in times of productive stagnation – engendering financial growth – because the forms that reproduce capital are distinct in character from capital's impersonal relations. Accordingly, this interpretation of *Capital* argues, what reproduces capital must be other to the logic of capital.

Chapter 2, framed by twentieth-century debates prompted by the long-term demise of the gold standard, will delve into the workings of the money form with the help of Suzanne de Brunhoff's spearheading study of money and its workings across Marx's three volumes of *Capital*. The purpose of the chapter is to explain the formal place of fictitious capital and its logical role within the movement of value as structurally, immanently other to capital. This will develop the argument that the logic of capital's reproduction can only be adequately grasped by a concept of reproduction determined by non-capitalist social relations as conditions of possibility. By sharpening our understanding of the nature of subjection brought about by the dialectical tension of interiority and exteriority within capital's social relations – and therefore of what connects multiple sites of social domination – the analysis will advance the tools necessary to identify possible pathways to the historical change of, and resistance to, such relations.

---

52  These non-capitalist relations that work to reproduce capital are not the same as 'extra-economic' forms used to sustain accumulation, expansion and exploitation of surplus value (as in the case of Nancy Fraser's depiction of the role of the extra-economic in social reproduction). Logically, economic forms, too, are engaged in reproduction as non-capitalist economic forms. This analysis differs from Fraser's, maintaining that there are non-capitalist economic forms internal to capital's abstract form that work as the preconditions for capital's reproduction. Furthermore, this analysis claims, we cannot unpair the extra-economic from the economic. Rather, the relation is more dialectical: The extra-economic is internal to the economic and the non-capitalist is internal to the capitalist, providing 'immanent externalities'.

CHAPTER 2

# Money Form

## Introduction

The monetary economy – as the medium of circulation of capital – offers a privileged vantage point to understand how financial capital and interest-bearing capital are implicated in the production and reproduction of capitalist societies. This chapter assumes this vantage point, focusing on the work of Suzanne de Brunhoff, a central figure behind the French Marxist Monetary School, whose foundational work *Marx on Money* is a novel account of the role of money across the three volumes of *Capital*. This chapter proposes a return to de Brunhoff's monetary theory both due to her clarification of the place of money in Marx and because her analysis offers an indispensable contribution for understanding the inner workings of capital today where markets are heavily financialised.

De Brunhoff's work establishes money as the key conceptual tool that allows us to understand, from a macro level, the relationship between lived experience in capitalist societies and the structural mechanisms that reproduce the accumulation of capital, especially in their financialised forms. De Brunhoff's analysis can be broadly understood as grounded in the conceptualisation of money as a commodity, and therefore also grounded in an account of abstract labour. Her account examines credit money's deviation from the commodity, departing from a monetary theory of credit. By grounding money in commodity money, de Brunhoff understands the money form as a distinct formal appearance of value within the value-form relation – which consists of money, commodities, and capital. Within this set of relations, money's character as a medium of circulation facilitates value's passage between forms to reproduce the conditions of production and to accumulate more value.[1] At the same time, de Brunhoff insists on money's simultaneous deviation from the commodity, claiming that as a general equivalent, money is an independent variable within the value form. According to de Brunhoff, money's independence is rooted in money's threefold character, at once a measure of value, a medium of circulation and a general equivalent; these three characteristics together produce money as a distinct social form, or 'money as money'.

---

1   De Brunhoff's reading is not antithetical to a value-form analysis.

Accordingly, de Brunhoff's framework reflects how we might understand the relationship between credit money and accumulation. The conceptualisation of this connection underpins de Brunhoff's claim for the necessity to understand 'money as money'. While money is a commodity when money circulates as commodity, and money is capital when money circulates as capital, money's independence is secured in its role as credit. The 'distinction between money as money and money as capital is nothing more than a difference in their form of circulation'.[2] Credit money is a form that finances production and takes place before valorisation, and in this way circulates as un-valorised value (value form that is 'not yet' value form). Therefore, money must also be understood as existing as a form in and of itself. As de Brunhoff claims,

> The Marxist theory of money interests us primarily because of its integration with the theory of the capitalist form of production. Since money is part of the machinery of capitalism, its role is determined by its function within the entire pattern of capitalist economic relations. According to Marx money is "a social relation of production"; therefore, under capitalism, it is part of the capitalist system of relations of production. But it participates in them in its special fashion, by existing in the form of money, and the monetary problem consists precisely in knowing the meaning of this strange existence as money, inseparable but distinct from the other relations characteristic of capitalism.[3]

De Brunhoff's contention is that, while money is derived from the commodity form, it cannot be reduced to any other variable within the capitalist system; it therefore needs to be theorised from the point of view of its specificity as an independent variable.

De Brunhoff's intervention, developed through a reading of Marx to the letter, is crucial because there is a general lack of specificity in theorisations of money's function elsewhere.[4] This causes difficulties for mobilising the political consequences of differing analytical commitments. Money is often straightforwardly assumed to be a commodity form or, opposingly, taken for granted as an archaic or transhistorical form with its history in pre-capitalist

---

2  Marx 1990, p. 247.
3  de Brunhoff 2015, p. 19.
4  In Marxism, money has been understood from a variety of interconnected and competing perspectives: from a precapitalist form (maintaining its precapitalist functions and put to use within capitalism as appropriated from the past); as a means to control the conditions of production (or political command); as a commodity; and/or as a value form.

market economies that have been appropriated for capitalist purposes. The latter is effectively money as understood from the point of view of a realist bias,[5] as in the case of classical and neo-classical economic theory. In both cases, the commodity form receives disproportionate attention over the money form as the constitutive factor in the capitalist mode of production. Focus on the commodity without attention to its dependence on money renders the political stakes of credit and finance unclear when the money form is a central mediator of social domination. Credit and finance are developments of the money form, which is both a commodity form and a general form set apart from the commodity form. The relationship between money and the commodity is hence both a complex and significant aspect of a critique of political economy, and it requires further attention if one is to accurately theorise subjection within capitalism, where credit and finance play significant roles.

When acting as credit money, the money form is not valorised value and therefore cannot be understood simply as a reified form (imposing impersonal domination). Due to a disproportionate focus on *Capital* Volume I, where readers find a sustained analysis of commodity fetishism, there are a number of difficulties within existing Marxian literature on the relationship between money and credit. In Volume I, the analysis of money focuses on its roots in the commodity, which, as such, operates as a formal expression of value. Most of the literature that only addresses Volume I has, therefore, produced a narrow focus, blinding Marxism to the role money plays as a variable in reproducing capitalist social relations. I contend that the money form cannot be understood without a reading that spans all three volumes of *Capital*, where money's role is conceived as a medium of circulation facilitating the schemas of reproduction in Volume II, and where money's role is developed as a medium of debt and finance in Volume III. These are two points that cannot be straightforwardly derived from an analysis of the commodity.

As de Brunhoff emphasises, while money has its genesis in the commodity, it is also necessarily independent from the commodity and therefore produces different formal results, such as credit money and 'fictitious capital', which have come to 'evade the conditions of the circulation of capital'[6] and thus commodity circulation and production. Following de Brunhoff, the presentation of money in *Capital* requires an address of how Marx's elaborations in Volume I work in conjunction with Volume II and Volume III.

---

5  Money is treated this way within post-Marxist accounts that treat the elimination of the gold standard to result in the elimination of money's commodity form.
6  de Brunhoff 2015, p. 94.

The first part of this chapter, 2.1, 'Political Subjectivity and the Monetary Link between Italian Operaismo and Capital Logic', highlights the significance of a debate between de Brunhoff and Lapo Berti in *Primo Maggio*. In doing so, I focus on the political consequences of de Brunhoff's monetary theory, which I then apply in the second section, 'Money as Money'. This is done by first sketching out the conjuncture to which de Brunhoff reacts: the 1974 Monetary Crisis – a consequence of the eventual elimination of the convertibility of the American dollar into gold – and the state-credit monetary systems that replaced it. This historical change reconfigured the way in which Marxists considered money's relationship to the commodity form and the consequences thereof. This section sketches out the details of what was found to be at stake in 1974, arguing that, theoretically, a bifurcation occurs within western Marxism between Marxism (a strand that retains the labour theory of value) and post-Marxism (a rejection of the central persistence of the labour theory of value). As a result, subjection, and therefore possible political action, have been theorised in very different ways. In this section, I argue that an analysis that draws on 'money as money' can account for both Marxist and post-Marxist concerns.

Through the explication of de Brunhoff's formulation of 'money as money' I differentiate the formulation from proximate conceptualisations of money. The practical application of ideas played out in this section are discussed to illustrate the political stakes behind the interpretation of *Capital* where capital's reproduction is argued to be a system of monetary reproduction. The debate interpreted here, between de Brunhoff and members of the *Primo Maggio* collective, is one of the few theoretically detailed and politically applied engagements featuring different possible interpretations of the money form in Marxism. The debate represents a turning point historically, addressing shifts that have produced the framework for financialisation in its current iteration, allowing us the means to answer the still relevant question: how would a Marxian analysis deal with the movement from a commodity-money system to a state-credit monetary system? Hence, an interlude into this debate is instructive for grasping why the orientation 'money as money' is not only politically significant but a correct interpretation for grasping capital's reproduction. This section exemplifies why a 'money as money' approach forms the method of analysis of capital underpinning this book.

The second section, *2.2 Money as Money*, establishes an interpretation of money derived from de Brunhoff's monetary theory. Here, I emphasise the need to ground analysis of capitalist social relations in an understanding of money as a general equivalent and as an independent variable. I argue that the reproduction of capital is a system of monetary reproduction because money both opens and closes the system's cycle. Subsequently, I show that the institution that reproduces capital must act as a supplement to the logic of capital, albeit as one that is immanent to capital. Money must therefore be both capitalist and non-capitalist: 'money as money'. A 'money as money' interpretation is proposed to reflect how money exceeds its role as a form of value's appearance, such as in its appearance as credit money. Credit money mediates non-capitalist social forms, which act as conditions of possibility for capital's forms. This interpretation is a logical foundation for a broader social theory of reproduction established in Chapter 5, 'Marx's Social Theory of Reproduction', where mediums of reproduction of capital are articulated as 'immanent externalities'. Such immanent externalities include the money form, human life and nature.

An emphasis on 'money as money' will prove necessary to gain insight into the dominance of finance and the significance of this domination for production and social reproduction. I deploy de Brunhoff's work as a logical tool to analyse subjection in capitalist social relations, considering both theories of interpersonal subjection and subjectivation – as developed in post-Marxist analysis – and a value-theoretical approach that exposits the nature of impersonal forms of domination. I argue that these two theoretical orientations can be synthesised. By bringing the two orientations, Marxist and post-Marxist, together through a monetary theory of capital's reproduction, a stronger interpretation of subjection within capitalism can emerge.

## 2.1 Political Subjectivity and the Monetary Link between Italian Operaismo and Capital Logic

### 2.1.1 *1974 Monetary Crisis*

De Brunhoff's work can be situated at the centre of debates regarding the implications of the 'monetary crisis' of 1974, when the dollar as 'currency of currency' was thrown into question not very long after the 1971 suspension of the convertibility of the dollar into gold. This suspension of convertibility was combined with a universally adopted 'state-credit monetary system', where money and monetary policy had 'become an important terrain of class struggle'.[7] These historical changes facilitated exceptional financial growth, leading to substantial changes in labour relations and the distribution of wealth, which continue to determine debates surrounding financial capital in the present. Intervening

---

7  See de Brunhoff and Foley 2007.

primarily in Keynesian discussions, and strongly standing up for a Marxist economic program, de Brunhoff's theoretical developments cannot be separated from her focus on practical social change.

De Brunhoff's commitment to developing a theory with the purpose of fostering practical social effects is evident when looking at the intervention she made within the pages of the workerist journal, *Primo Maggio*, which triggered a debate between herself and Lapo Berti, who was a representative of a research program on 'money' initiated by members of the journal's collective. De Brunhoff's novel intervention in *Primo Maggio* reflects a bifurcation in western Marxist theory between two almost completely disconnected sides. It is crucial to reflect on this debate in the present, as the theoretical presuppositions that were being worked out at the time still mark competing perspectives between Marxist and post-Marxist theory today. The divide between de Brunhoff and Berti largely rests on a distinction between two readings of the role of money: money understood as a means to gain command over exploited labour (or 'money as capital') versus money as the commodification of social life, considered through the commodity-money-capital relation. These two theoretical positions continue to divide the field. De Brunhoff's focus on 'money as money' offers us a unique position from which to account for concerns propagated between these two sides and, in doing so, provides a unique entry point to understand the internal complexities of the capital relation in the present, where there are multifaceted, conflicting, contradictory and co-dependent forms of subjection derived both from interpersonal and impersonal forms of domination.

### 2.1.2   *New Laws for Action*

Within the Marxist tradition, theories of subject formation understood in relation to historical changes in capitalist modes of production from the post-war period to the present reflect a pervasive reorientation from the commodity form to the money form as a primary locus of social organisation.[8] This shift to a

---

[8] As we will see, in the Italian context this includes the Operaismo tradition, which came to understand money as a form of political command and linguistic determination rather than a value-form relation (always understood as and/or in relation to the commodity form). This is best exemplified in the work of Christian Marazzi. A focus on the deregulation of the gold standard, and attention to 'financialisation' and credit more generally (especially since the financial crash of the early twenty-first century), has influenced disparate works by writers proximate to the Marxist tradition, including Wolfgang Streeck, Mark Fisher, David Graeber, Thomas Piketty, Augusto Graziani, David Harvey, Giovanni Arrighi, Costas Lapavitsas and Richard Wolff, to name only a few. This focus is ubiquitous and vastly internally differentiated.

focus on monetary theory in Marxist discourse (broadly speaking), with its aim to interpret the phenomenon of deindustrialisation in the Global North, has experienced a tumultuous relationship with the Marxist tradition from which it develops.

Much Marxist theory has tended to see the abandonment of the gold standard as an abandonment of its commodity form and has subsequently examined money through a realist lens. This undermines the fundamental dynamic of Marx's critique: the contradiction between essence and appearance internal to capitalist social form, which contains a dialectical relationship to its material referent. However, this misappropriation is far from a glaring misreading; rather, it reflects an internal differentiation within Marx. In particular, this 'realist' or empirical account of the money form (where money is seen to directly represent value rather than to be a form of appearance of value) is symptomatic of the very structure of the money form within Marx's critique, which, as a medium of circulation, exists as a generality that is other to capitalist forms.[9] This is why post-Marxist accounts can reject the labour theory of value and still theorise money from a perspective that does not stray from aspects of Marx's theorisation of the role of the money form (especially in its iteration as credit money). This is true of the literature developed within the Italian Operaismo tradition.[10]

Through the elucidation of the money form, theorists within the Operaismo tradition came to establish a post-Marxist discourse (based on a rejection of the labour theory of value), leading to some of the most important work on subjectivity and the social relations produced by financial capital post-

---

9   See de Brunhoff 2015.
10  The Operaismo tradition, also referred to as 'workerism', is a body of Marxist literature coming out of Italy, largely in rebellion from the standard theoretical line of the Italian communist party. Operaismo, associated with the writings of Mario Tronti and Antonio Negri, reads mid-twenty-first century capitalism in Italy from the point of view of workers' struggles, where the working class is understood as active and capital as reactive. The movement set out to read Marx as a tool to understand current conditions, leading to the establishment of a 'post-Marxist' line of thought. This was especially true of the way in which the literature was developed by subsequent thinkers associated with Autonomia, such as in Antonio Negri's later works and the writings of Maurizio Lazzarato, Paolo Virno, Franco "Bifo" Berardi and Christian Marazzi, who broadly rejected the social reality of the labour theory of value for other theories of subjectivation associated with governmentality and linguistics. Far from a unified theory, Operaismo reflects a broader encounter with political and social processes that base themselves on transformation, conflict and dissent as a means to understand the changing nature of capitalism in the post-war period. A common thread to this body of literature can be found in its specific attention to subjectivity without construing a unified political subject (Nigro 2018, pp. 173–5).

Bretton Woods.[11] However, much of the resultant work evinces an inability to grasp the money form in its full complexity – as independent from production and at the same time internal to production. This oversight led to the neglect of dynamics of rapid industrialisation (simultaneously accompanying western deindustrialisation) occurring in the Global South and the ongoing extraction of surplus value at the heart of the capital relation which structurally sustains the independence of financial operations. Such neglect led to failures within social movements, from the 1968 uprisings to the later Autonomist movements of the 1970s and anti-globalisation movements (all of which drew deeply on the Operaismo tradition). For the most part, political organisations were too strongly committed to the possibility of 'political command', believed to be possible in the money form. Thus, movements envisioned the possibility of a political subject with a particular kind of agency. In doing so, they underestimated the repressive power and persistence of the value form, an underestimation that corresponded to a theoretical failure to critique abstractions.

Nonetheless, Operaismo interpretations of the uncoupling of money and finance from the capital relation remain consonant with dynamics immanent to Marx's account. The problem is that Operaismo accounts can only ever be partial, missing the essential methodological charge bequeathed by the Marxist tradition – which is to understand the implicit and ontological co-dependency between production and circulation. Without this, analyses miss the fundamental insight of Marx's social theory: the economy is not an objective realm of social organisation but a totalising social form, a social relation of production that determines action independently of conscious engagement. The conditions of production that structure the movement of capital determine the value of the money form, not the other way around. In light of the Operaismo position, one might repeat Marx's rebuke to Proudhon (who thought that what was wrong with capital was the monetary system): 'the doctrine that proposes tricks of circulation as a way of, on the one hand, avoiding the violent character of these social changes and on the other, of making these changes appear to be not a presupposition but a gradual result of the transform-

---

11   In 1971 the United States terminated the US dollar's convertibility into gold and with that ended the Bretton Woods system. The result was that the dollar became fiat currency, which nonetheless continued to function as the standard of currency within the global banking system of the West. The Bretton Woods system itself existed on the basis of executing the decline of the gold standard. Since 1933, increasing demonetisation of currency was replaced by state debt, which was a trend further intensified by the 1944 Bretton Woods agreement (see de Brunhoff and Foley 2007).

ations in circulation'.[12] The lack of ontological engagement with the structural co-dependency between production and circulation therefore fundamentally misunderstands of the reality of capitalism.

By looking at the development of monetary theory and the corresponding political implications within Operaismo – subsequently applying a robust theorisation of the money form thereto – analysis can remobilise meaningful insights from this tradition. With a strengthened understanding of the money form, analysis can acquire better purchase on the political theorisation developed by Operaismo, potentially surmounting the limits encountered in a one-sided interpretation of the money form. In approaching an analysis of capitalist social relations from the point of view of the class relation rather than production, the Operaismo tradition became well known for departing from many of Marx's formal presuppositions. Concerned with the inability of orthodox Marxism to represent contemporary life, the collection of theoretical contributions emerging from this tradition privileged the attempt to understand the changing dynamic of labour taking place in the Italian post-war period. The changing dynamic was addressed through the deployment of the term 'class composition', used to understand the re-composition of labour and class under different societal formations. This showed how historically specific forms of class formation are implicated differently in the potential destruction of the capital-labour relationship. These changes in class composition were understood as not merely resulting from shifts in production but as influenced by broader changes in the governance of society: from forms of monetary circulation to technological change and consumption. The thematic of 'class composition' was used as a means to identify what Mario Tronti referred to as 'laws of development'[13] and 'new laws for action'[14] in a theoretical approach derived from a particular understanding of the development of capital and the corresponding actions that could be taken thereagainst. The point was to address the way in which actions will carry different meanings based on historically specific class formations, or 'compositions', as dictated by production and consumption as well as the state and the bank.

Characterised by attention to the assembly line, automation and the role of finance in production, Operaismo was unified[15] by the insight that industrialisation engenders mass production. This insight went hand in hand with

---

12    Marx 1973, p. 122.
13    Tronti 1971, p. 89.
14    Tronti 1971, p. 15.
15    The multiple interpretations associated with this movement vary from monetary circuit theory through Tronti's political ontology (i.e. *Workers and Capital*) to Negri's Spinozism.

the phenomenon of mass consumption. That is, since commodities are made cheaply and sold cheaply *en masse* by a large population of workers, the workers themselves are buying these commodities; therefore, the very consumption of society as a whole, including that of the working classes, is fuelling the need for objects produced in the factory. The result is increasing automation – producing menial roles to be carried out by workers – combined with mass consumption. Consequently, it was theorised, the realm of the factory has entered the realm of society, while both sides (the factory and society) were seen to be increasingly devoid of human meaning and sentiment. In Operaismo theory, this dynamic was referred to as 'Fordism', reflecting the American variant of mass production and consumption.[16]

Italian Operaismo based its revolutionary project on the idea that Fordism was a form of social organisation wherein the collective political subject resided: the mass worker. This particular subject was seen to provide the subject position that could determine what Tronti refered to as 'new laws for action'. As Steve Wright observes in his reconstruction of the Operaismo project, Tronti proposed to identify laws of development 'through which the economic input labour-power periodically constituted itself as the political subject working class, able to challenge the power of capital – and ultimately, the operaisti hoped, the continued reproduction of the capital-relation itself'.[17] Tronti claimed that there were times when a particular class formation can be more or less able to constitute itself as a political subject, by which (to paraphrase) 'new laws of development' could then lead to 'new laws of action'. Because a given factory was made up of thousands of workers, trained with particular technological characteristics, the Operaismo theorists claimed that the mass-worker figure, who lived in a society that mirrored their production through mass consumption, contained a latent revolutionary subjectivity. Since the factory provided these workers with their means of subsistence and objects of use, the workers became interconnected within the system as both wage earners and as consumers. It is significant that in this particular form of labour, the structure of a worker's pay cheque was complex; they had both a base wage and a variable wage that was linked to their productivity, as

> … there were also items that corresponded to contractual gains like pace with inflation, family allowances, overtime, production bonuses, com-

---

16  This is a term that, problematically, is often used for the sake of historical periodisation rather than describing a particular labour relation that was specific to certain Western countries, cities and suburbs.
17  Wright 2014, p. 369.

pensation for night work and hazardous work, etc. ... [while] the organisation of Fordist production was not only the dominant system within the factory, but also projected its rigid structure onto society, onto urban and suburban mobility, housing settlements, shopping houses.[18]

The wage variable gives the mass worker a modicum of freedom. However, this freedom was formally reduced to the narrow forms of consumption permitted within Fordist societies. The form of consumption available to the mass worker mirrored the alienated form of work. Operaismo responded to this phenomenon by evaluating the ways in which this kind of social formation leaves the working population more susceptible to politicisation as an antagonistic subject. As Sergio Bologna recalls, 'thousands of workers left the factories early in the morning after working the night shift, while many others were already outside waiting at the gates to enter for the first morning shift, [and] this was the best moment to distribute and spread the flyers of *Classe Operaia* and *Potere Operaio*'.[19] Not only was this form of social organisation particularly open to the efficient distribution of knowledge on a grand scale, since mass amounts of people worked together in the same factories, but workers were not isolated from one another since they were accustomed to working collectively in the context of the factory. Therefore, this mass contained not only antagonistic subjectivity, as the class of labourers under capitalist social relations, but also both the socialised and intellectual elements necessary for politicisation. An awareness of this condition underpinned Tronti's theorisation of the 'autonomy of the political', an idea claiming that if mass workers collectively refuse work as their means of subsistence, they consequently become a collective political subject opposed to capital (and so an autonomous form). In rejecting work as the place where one claims a means for subsistence, this class becomes a political force, not merely an economic force; it is able to command the mode of production and circulation to different ends than that of capital. Importantly, Tronti argued that 'autonomy involved an historical process, in which the autonomy of the political emerges as a consequence of the fully accomplished rationalisation of the economic',[20] This is precisely what Operaismo saw in the mass worker: an accomplished rationalised form of social organisation endowed with the technological specificity that would enable them to become a political class. This political class presupposes an adequately ration-

---

18   See Bologna 2014.
19   *Classe Operaia* and *Potere Operaio* were the names of the journals containing the output of early workerist literature (See Bologna 2014).
20   Farris 2011, p. 44.

alised state apparatus, putting politics in command of social production and circulation rather than capital. Here, theoretical practice takes a point of view that 'requires one-sidedness in order to grasp the whole anteriority of the workers struggle to capitalist development'.[21]

### 2.1.3 Classe Operaia *Literature*

The journals *Quaderni Rossi* and later *Classe Operaia* were established to provide a means to circulate Operaismo literature, and came to facilitate a political and cultural foundation for a new leftist movement. This new movement broke from the dogmatic applications of Marxist thought seen on the official Italian left at the time, which, rather than analyse the empirical specificity of social life, clumsily imposed Marxist categories onto social life. Broadly, the aim of the journals was to 'develop political strategies based on theoretical assumptions and practical experiences out of the concrete struggles of the 1960s'.[22] With the first issue of *Quaderni Rossi* printed in 1961, and the final issue of *Classe Operaia* in 1967, Italian Operaismo literature was a the sum of these journals, which were often co-authored by Operaismo theoreticians and factory workers. Initiated by Raniero Panzieri, Mario Tronti, Alberto Asor Rosa, Massimo Cacciari and others, *Quaderni Rossi* was intended to analyse society through a collaboration between intellectuals and workers who, first and foremost, sought to understand the current empirical conditions of life from the point of view of the working class without over-determining such a position via the imposition of a ready-made Marxist position. The journal's editorial collective was made up of Weberian sociologists from Turin, who were interested in the dynamics of the changing industrial society and a group of young intellectuals around the Partito Comunista Italiano (PCI) in Rome, who rejected what they saw as a strong division between social research and political intervention.[23] The journal was founded with the purpose of creating non-sectarian work, responding to what the groups jointly considered to be 'a crisis in the relationship between the "conditions of struggle" and the "politics of the parties"'.[24] In 1963, following a conflict between members over how they thought political intervention in class struggle should be conceived, *Quaderni Rossi* fell apart, and *Classe Operaia* emerged. Headed by Tronti, *Classe Operaia* included contributions from Toni Negri, Romano Alquati and Alberto Asor Rosa, and aimed to relay tactical discussions that addressed changing power relations. Yet, the

---

21  See Filippini and Macchia 2012.
22  Filippini and Macchia 2012, p. 8.
23  Filippini and Macchia 2012, p. 10.
24  Ibid.

journal came to a decisive end in 1967 when it appeared to the group that their project was quickly turning into either 'a politics of mere survival or its recycling in sectarian experiences'.[25] It thus became clear to contributors that their ideas would not be realised in the mass form on which their theoretical stance relied.[26]

It was not until 1972 that another journal, *Primo Maggio*, appeared. In line with the Operaismo programme, the journal emerged within the editorial collective Calusca in Milan. The bookseller Primo Moroni – who opened the Libreria Calusca, which continues to be a landmark of the extra-parliamentary left – supported the publishing of the project, assisting in the acquisition of its relatively vast readership. The very first issue of the journal sold 1700 copies, the second 2300, circulating in universities, prisons and even amongst the top managers of the Italian Central Bank.[27] Significant contributors to the journal include Sergio Bologna, Lapo Berti and Christian Marazzi. Marking a new generation within the Operaismo tradition, *Primo Maggio* was initiated with a renewed project in mind, which was mainly to understand the relationship between money and labour within the context of a world that was undergoing rapid financialisation. While the method to achieve this entailed 'changing the social role of political intellectuals by innovating the methodology of historiography, sociology, economics and political science',[28] *Primo Maggio* paid specific attention to the Marxian analyses of money in relation to monetary disorder occurring in the 1970s, where 'the 1971 disconnection of dollar from gold and floating exchange rates opened the way to a fragmented international monetary system, and made accelerated inflation the new form of the crisis of overproduction'.[29] This renewed focus responded to the growing role of monetary intervention in Italian social life, manifest in proliferating financial operations and the destabilisation of the social securities of Fordist factory work.

In particular, the journal initiated a workgroup on money that responded to the elimination of the gold standard and the corresponding re-composition of class based on the standard's replacement with the fiat money of the American dollar.[30] In the Italian context, the result was increased unemployment combined with a sharp rise in prices. In order to try to find possible modes of

---

25  Filippini and Macchia 2012, p. 11.
26  Lucarelli 2013, pp. 30–50.
27  Lucarelli 2013, p. 1.
28  Lucarelli 2013, p. 1.
29  See Bellofiore 2016.
30  Fiat money is a credited form of money issued by the state.

action against this dynamic, the project of *Primo Maggio* was to understand the function of money in the capitalist system, in an attempt to 'understand money as a privileged tool through which capital might outmanoeuvre the workplace-unrest of the period'.[31] Here, the concepts of 'money capital'[32] and 'class composition' were central to the inquiry and, furthermore, were found to have a direct relationship with one another. This was, in its premise, nothing less than an articulation of a key presupposition for, and influential link to, the emergence of post-Keynesian monetary circuit theory that would later be developed in the work of Augusto Graziani in 1989.[33] As Graziani claims, the circulation of money 'does not solely exercise the function of permitting easier commercial relations, but also serves the much more relevant function of putting the class of capitalists in relation to the class of workers'.[34]

### 2.1.4 Money and Financialisation

The implementation of the workgroup on money as a subproject within the *Primo Maggio* editorial group is unsurprising, as the journal, from its inception, was directed towards problems of overproduction and inflation related to the monetary crisis. The first issue contained an article written by Sergio Bologna on the problem of money, entitled 'Money and Crisis: Marx as Correspondent of the "New York Daily Tribune"'. This article established the parameters for the project of *Primo Maggio*, arguing that the relationship between crisis and the money form 'provides the key to a re-interpretation of political institutions from the standpoint of monetary organisation, and of the laws of value seen from the viewpoint of a stage of capitalist development now in its maturity'.[35] By looking to Marx's analysis of the monetary crisis of 1857, Bologna problematised the relationship between *money as capital* (money that is used to acquire more money) and *money as commodity* (money as containing a direct link to labour), a differentiation posited as the presupposition for the monetary differences between the sphere of production and that of finance. By initially

---

31  Wright 2014, p. 371.
32  This refers to accumulated value containing surplus value, existing in the form of appearance of money that can be used to fund and therefore determine the means of production – and therefore also command over labour.
33  See Graziani's book *The Theory of the Monetary Circuit* (1989). Monetary circuit theory has been taken up in recent years by Riccardo Realfonzo, Giuseppe Fontana and Riccardo Bellofiore, as in Bellofiore's 2005 essay, 'The Monetary Aspects of the Capitalist Process in the Marxian System: An Investigation from the Point of View of the Theory of the Monetary Circuit' (Moseley 2005, pp. 124–39).
34  Graziani and Vale 1983, p. 22.
35  See Bologna 1973.

establishing the relation between monetary command and class composition, the workgroup on money took capitalism to be a monetary production economy.

The monetary production economy specifically pertains to the role of credit money, which accords with the presentation of credit money in Volume III of *Capital*. The group aimed to show how those who are able to command money through the mediation of credit control the reproduction of the system, where money is created endogenously within the financial sector (more money is created by means of money, which is the position taken by monetary circuit theory that places a heavy focus on Marx's formulation for fictitious capital, $M\text{-}M^1$). Taking the standpoint that monetary 'crisis is a result of political and institutional choices concerning the credit sphere', Bologna, following Marx on the function of the *Crédit Mobilier* in his writings for the *New York Daily Tribune* (at the time an almost unknown text),[36] named the crisis as 'a revolution from above'. In this, he claimed that crisis functions to interfere with class composition as a way to prevent and reduce potential class conflict. Thus, following Lucarelli, the analysis 'of money in these pages is linked to the new forms of capital organization, to the new bourgeois elites which were supplanting mercantile ones, to the new forms of governance which characterize the modern State, and, finally to class struggle'.[37]

The workgroup was based specifically on an analysis of the new forms of capitalist organisation emerging as a reaction to the monetary crisis of the 1970s. In this analysis, we find the seeds of the later development of Italian heterodox economics, seen in the shift to the idea that capitalism is a monetary production economy largely based on the function of credit. Their observation was that, increasingly, money is produced endogenously, or within the realm of finance through speculation on loans acting as investments, and not exogenously, in the realm of production. This tendency had recently intensified due to the deregulation of money from the gold standard, freeing the creation of money from gold in its commodity form as the point of reference for the monetary standard.[38] The theoretical implication of a monetary production society, drawn by the group, was that money began to act as a social institution and so as part of the governance of society. Accordingly, the *Primo Maggio* workgroup on money made the claim that money was in command of class composition.

---

36  See Berti et al. 2016.
37  Lucarelli 2013, p. 5.
38  This political account of money's role accords with what Foucault went on to call 'governmentality'.

Money as a key link to the thematic of class composition is deeply ingrained in the Operaismo tradition. As Wright points out, Tronti's later novel argument in *Operai e capitale*, claiming 'the secret to overcoming capital's rule lay in labour refusing its function as labour power',[39] carries significant implications for the role of money. What Tronti posits is that labour is the measure of value because the working class is a condition of capital's social relations. Mainly, then, money is central to how the law of value asserts itself and, therefore, if 'only labour through its own struggles can determine the value of labour, then any working-class offensive of sufficient magnitude against capitalist command could threaten to undermine both the accumulation-process and the regulatory mechanism upon which commodity-exchange is premised'.[40] This regulatory mechanism is the law of value, with value able to circulate between different commodity forms by way of money. Money becomes implicated in labour through wages, functioning as an index of the relations of force between capital and a new class composition led by the mass worker, 'the human appendage to the assembly line'.[41] Negri made a similar point when he claimed that the wage was the 'ultimate independent variable'[42] due to what Keynes referred to as the 'downward rigidity' of wages.[43] Therefore, the ultimate goal – to eliminate capitalist command over labour power – would be to uncouple the relationship between productivity and income.

### 2.1.5 *Money as Capital*

'Denaro come Capitale' ('Money as Capital') was the first of very few articles to represent the general position that the of the workgroup had on money; this was written up by Lapo Berti, appearing in the second issue of *Primo Maggio*.[44] Consonant with Operaismo's commitment to understanding money's subordination to labour within the production process, this article addresses four key problems: why it was that a new politics of money was required; the new role of money and monetary institutions in the class dynamics of the post-Bretton Woods world; the role of inflation; and, finally, the relationship between the rate of interest and the rate of profit.[45] In openly claiming not to have a 'systemic theoretical framework',[46] the observations that came out of the work-

---

39  Wright 2014, p. 371.
40  Ibid.
41  Baldi 1972, p. 11.
42  See Hardt and Negri 1994.
43  Hardt and Negri 1994, p. 44.
44  See Wright 2014, p. 381.
45  Ibid.
46  Berti 1974, p. 9.

group were intended to be provisional, with priority granted to their practical political function as a 'reference for action'. From the outset, the authors issued the caveat that in order to account for the 'overall unfolding of the crisis',[47] there would need to be a more substantial study undertaken on the matter. As Berti describes it himself, the text is 'conscious of the limits of its theoretical elaboration, [and] represented a kind of agenda, if not a manifesto'.[48] The approach of the text was twofold. On the one hand, the analysis aimed to understand the unfolding societal transformations, free from theoretical preconceptions. On the other hand, the text attempted to test a Marxist approach against empirical reality without bending reality to support a Marxist reading. These two sides were interconnected. Berti claimed that 'the new representation of the capitalist crisis was emerging from a close confrontation between Marxist standpoints and real processes, as well as ... instruments of analysis that belonged to the opposing field, that is, monetarism'.[49]

The article relied on the premise that the then-current crisis in monetary mechanisms, on a global scale, represented the crisis of capitalist command in the context of preceding relations of force. Berti and the workgroup considered the fall in the dollar to be the product of class conflict, commencing with a crisis in the hegemony of American fiat money. It was for this reason that they argued there were central aspects of Marx's theorisation that no longer applied to the circulation of money in its current function. Most essentially, they claimed that the category of 'commodity-money' no longer corresponded to capitalist reality in any immediate way. Instead, the predominance of 'the role of both national monetary institutions and international firms'[50] could be seen. According to Berti:

> ... The creation of money, with all the consequences that this process entails in terms of the distribution of income and the economy's equilibrium, is now a process that depends, in a theoretically unlimited measure, upon the decisions of the [national] central bank.[51]

The *modus operandi* of money was taken by the workgroup to have substantially changed, particularly in regard to the international monetary system, where there was no longer a fixed exchange rate against the gold standard or

---

47  Ibid.
48  Berti 1974, p. 18.
49  Berti 1974, p. 19.
50  Wright 2014, p. 382.
51  Ibid. Berti is quoted by Wright here.

any other measure of value.[52] In this regard, money was no longer a money commodity; rather, money functioned only as capital (accumulated value) that could be transformed into productive capital when lent to purchase the workforce and other means of production. Furthermore, it was thought that monetary policy had surmounted its previous boundaries, making more room for money to be instrumentalised for political ends. If money was free from direct or indirect conversion to a commodity standard, money could then be rendered a manoeuvrable variable – or an instrument of governance. Berti claimed that 'money had become an institution with a high political value'.[53] If money could be transformed into a governing instrument, monetary policy would then play a direct role in the struggle between classes. This is the central point that the workgroup aimed to uphold. To the workgroup, money could seem 'like a political practice consisting of a redefinition of the laws governing money circulation. As a result, money cannot be an exclusive, exact representation of wealth; rather, it must be conceived as regulated money, both at a national and at an international level'.[54] In this way, Berti argued, money ought to be understood as a command before a measure; money would be thus dislocated from its role as a formal bearer of value in the *automatic* functioning of capital.[55]

Later, Berti undertook a case study aimed at showing how the governing nature of monetary flows was not a technical task but a political one, presented in the essay 'Inflazione e recessione: la politica della Banca d'Italia (1969–1974)' ('Inflation and Recession: The Policies of Banca d'Italia [1969–1974]'). This text represented money not as a neutral bearer of value but as something 'maneuverable and maneuvered as an instrument in the repartition of revenue between wages and profits'.[56] Central banks would therefore have the power to decide on the amount of money that would enter the system, granting them the leverage to 'interfere in the level of relative prices of, for example, goods and labor, and so this leverage could shift social balances and alleviate the pressure of wage claims'.[57] Central banks retained a flexibility in order to coax national economic growth through the juggling of interest rates, money supply and exchange rates that were ever in flux.[58] The case study insisted that

---

52  Berti 1974, pp. 9–18.
53  Berti 1974, p. 19.
54  Lucarelli 2013, p. 10.
55  Capital, when understood as reproducing itself on the basis of the movement of the value form (and therefore on the basis of the labour theory of value), is seen to circulate 'automatically', imposing its movement on human bearers of the form.
56  Berti 1974, p. 20.
57  Berti 1974, p. 21.
58  Wright 2014, p. 382.

when the central bank loosened its credit flows, it was to allow industries to borrow for the purpose of financing a reorganisation of the workplace, where production was to be decentralised and new technology deployed. It is Berti's contention that this loosening of credit was a reaction to capital's inability to continue to hold command over labour within the production process, and that this was compensated by revolutionising capital's form of command to reinstitute control. Since the main problem for Italian capital was its lack of ability to command labour, it reacted by using credit to effectively exert governance based on a formal reconstitution of the role of money. In this, there was a shift in wealth on the side of the capitalist class, who transitioned from creditors to debtors, representing what Wright refers to as 'a new order of capitalist power'.[59] The submission of labour to capital entailed, according to Berti, a change in the balance of power.

To follow the workgroup's position to its logical conclusion, as Christian Marazzi (another of its key contributors) did, is to find that once money is no longer convertible into gold, it is no longer able to act as a general equivalent. The conceptual addition made by Marazzi was that if money is no longer a representation of commodity form (gold), then it is reduced to being a mere 'money-sign'.[60] This means that money does not represent the measure of value; rather, it becomes a linguistic sign that indicates a value, standing in for itself, which is incompatible with the standard Marxist framework of the commodity form. It is then further argued by Marazzi that because money is unpegged from gold, it no longer has the capacity to be a general equivalent against other commodities, since this would require money to take on the commodity form. Because it is the ability of a variable to act as a commodity that facilitates its capacity as a measure of value, money can no longer be such a measure. Yet, Berti et al. insist that this so-called 'money-sign' is one that ultimately represents money capital, a form that still exerts control over labour as an instrument for the production of surplus value. So, while money is no longer seen as a commodity, it nonetheless signifies social power and has the ability to exert domination over labour in the form of circuit theory.

A monetary theory developed from the 'money as capital position' (currently enjoying a revival through Modern Monetary Theory) emerged as monetary circuit theory. Much like in de Brunhoff's theory of 'money as money,' money (M) in monetary circuit theory is understood to open the circuit of commodity (C) production (M-C-M). Money both sets the economic circuit in motion

---

59   Wright 2014, p. 382.
60   Marazzi 2015, p. 43.

and closes it with the realisation of profit. However, unlike de Brunhoff's theory, monetary circuit theory is grounded 'in the concept of money created *ex nihilo*' and circulated through wages.[61] That is, capitalists pay wages with the money that they have borrowed from national banks. These banks borrow that money from the central bank. Money created 'ex nihilo' (i.e. out of nothing) is credit money that is only repaid after the production process takes place. In this dynamic, money is not required to pre-exist the payment of wages. Therefore, money is created based on the separation of capital and labour rather than their relation as commodity forms via the extraction of surplus value. While this reading is true of credit operations in a classical Marxist sense, it attributes the workings of credit money to the whole of the money form. The consequence of this position is that there is no longer a link to the measurability of value and its monetary equivalent. If money is not a commodity, then it is also not a general equivalent. This is because it is money's commodity form that enables the money form to put other commodity forms in relation to each other through the measure of value. Money's commodity form enables it to be a medium of circulation and facilitates money's nature as a general equivalent. Analytical problems stemming from the 'post-Marxist' 'money as capital' approach persist due to the focus on money as created *ex nihilo*, which is but one aspect of a broader process. De Brunhoff's intervention explains why, despite historical changes in money's circulation, the deployment of a capital logic approach is still required.

### 2.1.6 Capital Logic Critique

For a 1975 conference at the Feltrinelli Foundation, Sergio Bologna and Suzanne de Brunhoff both contributed to the seminar 'The Marxist Discourse on Money in Light of the Monetary Crisis'.[62] This occasion triggered a rare dialogue on the then-current monetary crisis from Italian Operaismo and more-classical Marxist positions. De Brunhoff maintained a commitment to Marxist understandings of the value form and 'money as money' in her reading of money for a paper that was staunchly critical of the theoretical presupposition of 'money as capital', as per Berti's presentation of the workgroup's theories. De Brunhoff maintained a commitment to Marxian understandings of the value form and 'money as money'. This presentation was published in the sixth issue of *Primo Maggio* and was accompanied by a response from Berti. De Brunhoff began by locating the monetary crisis as an acceleration of the rise in prices and depreci-

---

61  Marazzi 2015, p. 43.
62  Berti 1974, p. 22.

ation of currencies, which coincided with a weak dollar on exchange markets. These elements converged in 1974, at which point the crisis was identified by a 10% inflation rate in wealthier capitalist countries (bar West Germany). At the same time, the dollar came under scrutiny in respect to its function as the 'currency of currencies'.[63]

This crisis had prompted two different readings from Marxist theorists: a value-form reading on the one hand and a class-opposition reading with the premise of money as capital on the other. According to de Brunhoff, while these responses are disparate, they are nonetheless based on a common central idea: that, while the monetary crisis illuminates the specific role of money in capitalist societies when an increased amount of money in circulation is credit money, credit money's function cannot be uncoupled from the formal role of money in the realm of production. The point of view of 'money as capital' is based on a method that requires the reintegration of concepts used in bourgeois political economy that fail to grasp the complex and multifaceted nature of money. Such a misunderstanding, de Brunhoff argued, undermines the legitimacy of the implications drawn from the workgroup's reading – significantly, in its understanding of money as an instrument of command over labour.

The key to the disagreement between Berti and de Brunhoff lies in their differing positions on what actually constitutes money. These differing conceptualisations become visible when considering what happens to the relationship between value and labour with the abolition of the gold standard. How would a Marxian analysis deal with the movement from a commodity-money system to a state-credit monetary system? At stake in this are the significant implications of the understanding of the role money plays in relation to production, and the understanding of the political implications of the role of money more generally. In reaction to the elimination of the gold standard, monetary economics has largely developed through the rejection of theories, developed by Marx, concerning the relationship between the value of money and the money-commodity (predicated on the extraction of surplus value from labour).[64] This is precisely the line that Berti et al. followed: the replacement of the gold standard by state debt (the American dollar), serving as both a national unit of account and means of circulation, means that there is no longer a commodity, represented by labour time, working as the link between labour and the accumulation of value. For de Brunhoff, however, because the role of money as commodity put into place a historical social form, the elimination of the gold

---

63   See de Brunhoff 1979.
64   de Brunhoff 1979, p. 202.

standard does not change the formal dynamics of said form. Since money's origins are in the gold standard (money as commodity), this brings into being its formal structure, which nonetheless continues to define money regardless of whether the gold standard obtains or not.

In contrast, Berti essentially argued that the elimination of the gold standard resulted in the complete autonomy of money from the production process, allowing money to act as a governing institution. Rather than having been extracted from labour, money exerts command over labour. This not only inverts the dynamic behind the tension between capital and labour, it also abolishes Marx's labour theory of value. While both sides retain a commitment to the analysis of money as playing a role in production that has a certain form of autonomy, de Brunhoff's argument instead held the position that finance capital functions autonomously only within an immanent relationship to the value form, remaining dependent on the extraction of surplus value from labour. From this point of view, finance, although appearing to be independent from the production process, especially with the destruction of the gold standard, is always still a product of surplus value first extracted from the production process. For Marx, any value created by finance itself is purely fictitious, regardless of the income generated through interest. While fictitious capital creates many new forms of income (well-articulated by Berti et al.), this should not be confused with autonomy from the production process, as production and finance have always existed in relation to one another.

Finance is not inherently a capitalist form of economics: it only becomes a part of a capitalist system when operating in relation to a process of production that accumulates value based on the exploitation of labour and the extraction of surplus value. Finance, a form of money management, becomes an agent of capitalist production when put in relation to capitalist production.[65] When put in relation to capitalist production, money capital becomes the source of the credit system.[66] Interest from credit issued to facilitate the growth of the means of production is derived from the profit of production (surplus value). Money capital is the source of credit (fictitious capital) that then derives an income for the owner of money capital through obtaining interest, which is a share of the profits extracted in the production process. Therefore, finance only becomes capitalist when coordinated with the management of money capital in the context of its affiliation with relations of production. Money capital as the source of the credit system in the capital relation causes a split in the form

---

65  In just the same way that money has multiple histories, one capitalist and others that are non-capitalist, so too does finance.
66  de Brunhoff and Foley 2007, p. 199.

of capitalist accumulation: 'capital property outside the production process' and 'capital in the production process'.[67] This results in a division between finance and production (qualitatively different capital incomes). What this means is that although finance might characterise the historically specific mediation of capital, as emphasised by de Brunhoff, 'this in turn rests on the appropriation of surplus value from the exploitation of wage labour, which, however, is obscured by emergence of different forms of income, for example, interest and profit of enterprise'.[68] In contrast, Berti maintained that these new forms of income, rather than obscuring how we might understand the problem – generating an appearance that covers up the real relation – directly demonstrates how money functions as capital. De Brunhoff's position countered Berti by emphasising how the extraction of surplus value is obscured by these new forms rather than directly reflecting realist phenomena. Therefore, the basic relationship between capital and labour remains at work in the case of new forms of wage, yet hidden behind appearances of money as fictitious capital.

The theoretical dispute between Berti and de Brunhoff encompassed not only the question of the independence of finance, in the case of renewed social organisation of production and consumption, but also the question of changes in monetary governance.[69] From a Marxian point of view, it is only if money can no longer be said to be the universal equivalent that labour can be freed from its value-form relation, resulting in the accumulation of value as autonomous from labour (and hence the separation of finance from production). It is on this basis that *Primo Maggio*'s workgroup claimed that money itself becomes a governing force, rather than a formal mechanism or universal equivalent facilitating the passage of value. This position was informed by passages within the *Grundrisse*, in which Marx referred to money as a governing force in the form of a 'top-down revolution'. The dislocation of labour from value is revealed not from the point of view of workers' struggle, but from the point of view of the state-supported administrative function of the banks, which subsequently redistribute labour for their own means. In this case, money comes to acquire command over labour.

---

67  Marx 1991, p. 375.
68  de Brunhoff 2007, pp. 199–200.
69  The elimination of the gold standard brought with it other forms of institutional deregulation. The way in which production is financed has vastly changed; there is an increasing internationalisation of productive relations as well as an increasing amount of fictitious capital controlled by banks and bankers, essentially giving the banking system heightened institutional control over the financing of production and property ownership. This change is also a result of technological development surrounding the way in which money is invested, traded and lent through speculation.

To develop the standpoint that money takes command over labour as a governing form, the workgroup drew heavily from Bologna's retrieval of Marx's *New York Daily Tribune* work. Marx showed how Louis-Napoleon Bonaparte used the distribution of credit (socialisation of credit) to prevent class struggle during the crisis of 1857 through the creation of a bourgeois class; in operative terms, this was done by using credit money to command class composition. Seen by Marx as an operation of socialisation, we find a communism of capital by way of banking credit, or the creation of new money by a collaboration between the central bank and the state. Marx, paying attention to the contradiction inherent in this dynamic, also showed how the distribution of credit led to the co-optation of the working class by returning them to work during periods of crisis. According to Marazzi, 'this is what he would develop later in the third book of *Capital*: the capacity of capital to expand and promote growth through the forced socialization of credit'.[70] This historical analogy became important for the workgroup on money, especially the institutional dimension that relays the interrelation between the state and the banking system, and their role in controlling class composition. Clearly, this points to a productive way of understanding how the socialisation of credit reinscribes class relations during times of economic depletion rather than performing an emancipatory role. Focusing on the socialisation of credit without acknowledging its implicit contradiction – money's stronghold as a commodity form relating to the sphere of production – results in a failure to engage with the social form upholding this particular mechanism. It is only by relying on a conception of the dollar crisis as a development that excludes the relations of production that Berti could establish a direct link between credit and class struggle. The exclusion of production from the equation presented a one-sided focus that could not understand why redistribution (i.e. the socialisation of credit) cannot ultimately undermine the global persistence of the capital labour relation as social form.

Moreover, the exclusion of production from Berti's hypothesis did not allow us to explain capital when it appears in its monetary form. This form of capital allows for the purchase of labour power in a distinct way from credit money.[71] Such a distinction regards the essence of the forms as either derived from abstract labour in the former case (money as capital) or fictitious capital in the latter (credit money). The incapacity to generate a meaningful theoretical distinction between these two leads to further problems in the analysis. For example, if we identify credit as money capital, we lose the distinction between

---

70  Marazzi 2015, p. 41.
71  Credit money being a form that is not based on labour but on a relation to the central bank.

money, credit and money capital: these three distinct categories are central to understanding the way in which capital circulates. The specificity of the respective formal roles further lays the ground for the specificity of the labour relation within the larger dynamic of circulation. At the same time, within circulation, there are forms that, although reliant on surplus value extracted from labour, are also set apart from labour due to their ability to create fictitious capital with credit. To ignore these formal distinctions is to also ignore Marx's critique more generally and to rely on the use of concepts of money used in the tradition of classical economics. Due to their lack of criticality, such classical economic conceptions inevitably inhibit comprehension of the social relationships mediated by money and value forms. De Brunhoff, in response to Berti, emphasised that 'the presentation of ideas here is confused. It does not explain money in its form as capital that allows for the purchase of labour power, nor does it reflect the capitalist function of money which supposes the capitalist as a buyer, and the seller as the seller of a wage; that is to say it does not reflect a class relation'.[72]

When we identify money with credit, we miss the interworking of the social conditions that allow a use value to become money. Hence, we miss the crux of the Marxist enterprise: to show how commodity production is based on the social validation of labour incorporated in the commodity. This labour, put into the creation of the commodity, provides the commodity with its social character. Furthermore, generalised social labour only becomes so when the commodity given a price that represents an amount of money and then sold. It is this generality (homogenous social labour), imposed on the commodity, that makes it exchangeable with money, since money represents the general equivalent, or the expression of the relative value that expresses all other values. That is, 'money appears here as a commodity whose material expresses the relative value of all other values as a general equivalent. The sale of a commodity for money gives the commodity a value containing the character of socialised labour'.[73] By abandoning the capital-logic understanding of money, the workgroup also abandoned a critical reading of political economy, where money as general equivalent represents social relations in the abstract. Significantly,

---

72   de Brunhoff 1979, p. 186. This translation is my own, the original French reads: 'cette presentation est confuse. Elle n'explique pas la forme argent du capital, qui permet l'achat de la force de travail, ni la function capitaliste de l'argent, qui suppose "dans l'acheteur un capitaliste et dans le vendeur un salarie," c'est-a-dire un rapport de classe'.

73   Ibid. The translation is my own; the original French reads: 'la monnaie apparait ici comme la merchandise dont lamatiere exprime la valeur relative de toutes les autres, comme un equivalent general. La vente contre monnaie a le caractere d'une sanction sociale tu travail prive'.

de Brunhoff called for the re-evaluation of the workgroup's conclusions in the light of a capital-logic reading accounting for the inclusion of production, combined with a critical application of economic categories. The workgroup's focus on the management of money by the bank and the state, which still offers a truly significant reading of the historical changes in the management of production and consumption on the part of finance, might regain its political potential if interpreted alongside a renewed commitment to the critical categories developed by Marx. An exhaustive account of the role of money within Marx's critique would render the synthesis of these seemingly disparate sides of Western Marxism possible.

*Primo Maggio*'s theoretical presuppositions ultimately rejected the interpretation that subjection in capitalist societies occurs through the value form. This put in practice a conceptual move that eliminates the Hegelian metaphysical basis of Marx's critique and so prompts a realist understanding of monetary functions. However, these theoretical presuppositions also return the political subject to a central position, whereby individual life engages in practice. The value-form tradition, by contrast, often fails to fully grasp or articulate the centrality of practice and politics. This can be attributed to the fact that – in the capital-logic reading – capital accumulation is not primarily understood from the point of view of class struggle or the political. Rather, capital accumulation is understood from the point of view of social form, as constituted by the value form. However, within a capital-logic perspective, we find an emphasis on money as an abstract expression of the value substance (and the resulting centrality of commodity exchange), and this largely ignores the social mediations that structure the analysis of value form.[74] Such an orientation often leaves out the lived implications (such as actions that are effects of structures or psychological effects and identity-based differences) of the subjects as defined by these forms. The resulting separation of individual life from subjection to form results in an evacuation of the form itself; without the individual subject, form loses its meaning.

There are two quite different understandings of subjectivity and subjection internal to these respective interpretations that are not mutually exclusive in Marx. Through a robust account of the money form, we can provide a central link between the competing perspectives: abstract self-movement of the value form on the one hand, and the political subject produced in relation to the circulation of credit money on the other. Attention to the range of complexities internal to the money form engenders an ability to retrieve the central insights

---

74   Saad-Filho 2002, p. 29.

of both sides. This includes the breadth of analysis of forms of income and work in the context of historically specific financial forms offered by the Operaismo tradition, and the ontological foundation contributed by the value form tradition. The synthesis of these two hitherto distinct sides may bring us much closer to an adequate analysis of the complexity of actual modes of subjection, begetting theoretical tools to reexamine the limitations and possibilities of political subjectivity in the present.

## 2.2 Money as Money

### 2.2.1 The Genesis of Money

In *Marx on Money*, originally published in 1967 – de Brunhoff begins an analysis of the role of money in Marx's *Capital* from the point of view of a general theory of money, or 'money as money'. De Brunhoff argues for a theory of money that understands it as a product of capitalist relations, albeit a product that contains its own independence. To think this, we have to understand money as both a product of, and at the same time separate from, capital relations. As de Brunhoff claims, if money is not understood in its generality as a form that functions separately from the capitalist mode of production, 'one becomes unable to see how the general laws of monetary circulation … [also apply to a] capitalist form of production where there is a special monetary circulation, that of credit'.[75] From de Brunhoff's perspective, money is considered to have its own 'monetary relationship' that is separate from the capitalist relation of production (as an antagonism of capital and labour). Money is but one form that is used to represent value and to circulate it among other forms. Therefore, money is a form of the 'phenomenal manifestation'[76] of value,[77] and the very mechanism by which capital is circulated and valorised. But money is not capital itself, playing a purely formal role upholding capitalist relations. Money is a bearer of value in circulation, turning into capital at the point of valorisation. Although capital might present itself as money, when it appears as such, it is no longer money qua money but money capital.

    Money is understood by de Brunhoff as an abstraction separate from capital. Yet it becomes capital when put under certain relations. Importantly, this does not mean that money is somehow tied to the pre-capitalist use of money. Money in the context of capital relations finds its genesis in the commodity

---

[75] de Brunhoff 2015, p. 20.
[76] Marx uses *Erscheinung* to refer to this phenomenal manifestation.
[77] Value achieves validity in the form of money.

and is therefore not dependent on the priority of pre-capitalist economies. In this way, money reflects a dynamic that functions to doubly separate it from a realist concept of history. In the capitalist mode of production, money derives its history from the commodity – an abstract form immanent to capital – and, as such, money is endogenous to capital. Money under capital is a derivative of the commodity and accordingly functions to produce different social relations than pre-capitalist relations, playing a different social role with a distinct economic status. Therefore, in Marx, the genesis of money is not historical but conceptual: Money emerges as 'the expression of value contained in the value-relation of commodities from its simplest, almost imperceptible outline to the dazzling money-form'.[78] The simple form of value referred to here is the commodity's expression of value by way of another commodity.

This role of money is a result of its appropriation from being a thing of economic use value to standing in as *the form under which value becomes exchange value*. Money is able to do this through its distinct role as the universal exchange value; value can be exchanged only when value is formalised under a universal measure that can then account for differences in value. As Fred Moseley correctly states, 'in order for each commodity to be exchangeable with all other commodities, the value of each commodity must be comparable with the value of all other commodities in some objective, socially recognizable form'.[79] However, the measure of value is but one function of money that is entirely dependent on money's other function: circulation. As a medium of circulation, money provides the role of the means for realising social relations as both 'a means to set into motion the productive process',[80] and also through financing production and connecting the different moments and actors within the production process. For example, as noted by Augusto Graziani, since wages are paid with money, it is necessary for money to initiate production, in a process that concludes itself through the sale of commodities that then retroactively fund the financing of the wages, though only after the money that financed production has changed forms. Hence, within the production process, money must consistently be converted into other value forms through circulation. Money thus functions as the connecting agent behind social relations, as a medium of circulation. What de Brunhoff emphasises, in this regard, is that the circulation of money facilitates the reproduction of capital, *making the reproduction of capital a system of monetary reproduction*, as money both opens and closes the cycle.

---

[78]  Marx 1990, p. 139.
[79]  R. Bellofiore and Nicola Taylor ed. 2004, p. 148.
[80]  Graziani 1997, p. 27.

## 2.2.2  Money as Medium of Reproduction

To operate as a means of reproduction, money must exist as a form based on a combination of functions (medium of circulation, measure of value, instrument of hoarding) that together produce its autonomous nature. This gives money its own temporality, set apart from the temporalities that make up other social forms. Only through its own temporality can money connect the different temporalities internal to the social practices of production and social reproduction. This occurs through money's circulating function, which facilitates the passage of value between different forms. It is based on this that de Brunhoff describes the money form as an 'immanent externality' from the capitalist form, distinguishing between the relation of circulation and the relation of production.[81] De Brunhoff's use of the term 'immanent externality' refers to how money as form is not simply internal or external to the capitalist production and reproduction process. In order to reproduce the basic capital relation, money cannot be reducible to an expression of the capital relation itself. This makes the money form a form that is *other than* the capitalist relation of production, or a 'non-capitalist' form within capitalism. What de Brunhoff has in mind is an insistence that *the institution that reproduces capital must supplement capital's own logic.*

Circulation occurs both positively and negatively: positively in the case of exchange; negatively in the case of hoarding, where money stands still or is preserved in large sums in its own simple form, ultimately functioning, by acting as a reserve, to uphold money in its role as a general equivalent. While hoarding is theorised by de Brunhoff as a third function of money, this is a misunderstanding of circulation as only playing a positive role. Not circulating money, in the case of hoarding, is nonetheless an act of circulation, albeit through negation. Hoarding is negative circulation and is necessary to ensure money retains its character as a 'general equivalent', accounting for any such case where there is a demand for money in the form of 'hard cash'.[82] Hoarding upholds the rules of simple circulation (and therefore sustains its function as universal equivalent) through 'absorption and preserving the difference between the total money supply and money in circulation'.[83] In doing so, hoarding provides the

---

81  de Brunhoff 1978, pp. 120–1.
82  De Brunhoff uses the term 'hard cash' to describe what is commonly referred to as 'liquidity'. This is because 'liquidity' is a Keynesian term that represents a character of money linked to demand and investment not compatible with Marx's analysis of hoarding (see de Brunhoff and Foley 2007).
83  de Brunhoff 2015, p. 48. Note: De Brunhoff does not refer to hoarding as negative circulation. However, she emphasises the need to include hoarding as a necessary positive

monetary basis for credit and international transactions. Therefore, hoarding becomes implicated in the development of money from a universal equivalent and medium of exchange (as extracted from abstract labour) to the role of money as credit money. Credit money being a form of money that is both of the production process and independent from production, determining the financing of changes in the production process.

Not only is circulation necessary for the realisation of capitalist social relations, it also constitutes the nature of money, permitting its role as a measure of value. Money must go through the process of circulation. Therefore, we find that these two functions – circulation and measure – are mutually constitutive. This formal mutual constitution is something that occurs by way of money effectively becoming valorised (i.e. adding surplus value to its sum and adopting its social form) and therefore is a result of the product of commodity circulation. Hence, Marx's position 'that money as a commodity is therefore only a discovery for those who proceed from its finished shape in order to analyse it afterwards'.[84] Like Spinoza's one substance that must contain two attributes (thought and extension) in order to be one, for money to act as one in the form of the general equivalent, it must both function as the measure of value and the medium of circulation: an attribute defined by its motion.

### 2.1.3  *Money as Social Form*

Marx begins *Capital* Volume I with a commentary that articulates the function of commodities under capitalist relations. Here, the commodity is developed as a general social form and therefore, a relational category. The development of what is understood as commodities and money precedes any mention of capital (capital is not discussed until the fourth chapter), indicating the basic function of the two forms in the development of Marx's argument. However, the extent to which we find a logical and historical development beginning with these concepts (leading to the concept of capital), as correctly posed by Moishe Postone, 'must be understood as being retrospectively apparent rather than immanently necessary'.[85] To clarify, Marx does not intend this sequence to be historical; rather, it is logical and relational, as this mode of development is only possible when already within the capitalist social formation and therefore is not historically constitutive of capitalist relations.

---

structural role played by money, and not merely something that leads to crisis by causing inflation.
84  Marx 1991, p. 184.
85  Postone 1993, p. 127.

The development of money emerges in *Capital* Volume I only after Marx's initial discussion of the commodity, where he establishes that it is constituted by both a use value and a value (with the phenomenal form of an exchange value). The double nature of value indicates that every useful thing in a capitalist society can be perceived from the point of view of either quality or quantity. Within each point of view there are again many more properties, as things can be useful in various ways. According to Marx, the work of history is to find the 'manifold' uses of these things, or to extract from the manifold the form in which things come to be used.

The use of the word 'manifold' is significant in Marx's initial description of the commodity on the first page of *Capital*. He states,

> ... every useful thing, for example, iron, paper, etc., may be looked at from the two points of view of quality and quantity. Every useful thing is a whole composed of many properties; it can therefore be used in various ways. The discovery of these ways is hence of the manifold uses of things [*die manningfachen Gebrauchsweisen der Dinge*] is the work of history.[86]

In its Kantian use, the manifold refers to the pre-synthetical givenness of everything that is then represented through abstraction to produce knowledge of something. Hegel saw in Kant that the manifold entailed an act of subsumption, since for Kant particulars needed to be brought under categories in order to achieve representation. Marx's materialism reverses this mechanism by showing that the manifold, rather than being negated by the act of subsumption, is latently available to the work of history. When applying his observation to objects, the object of commodity (its use value) is the externality of the manifold that subsumes a greater 'given', which could be accessed in different ways depending on historical circumstances. For Marx, things (use values) have an 'intrinsic virtue', and what becomes externalised or imposed is the work of history as a concrete sociality. Therefore, Marx is philosophically anticipating the development of commodity as a form that is constituted by external, relational concepts, as the work of a spontaneous historical development that bears the many properties of each point of view (quality and quantity). This supports Postone's argument that there is no historical development of one form bringing about another: instead, there are forms that are mutually constitutive and driven by historical specificity to make certain qualities present and others latent.

---

86   Marx 1990, p. 125.

This appropriation of the manifold internalised in the commodity character – as applied to the uses of things or their use value (as the work of history) – is also applied by Marx to the 'socially recognized standards of measurement for the quantities of these useful objects'.[87] This suggests that exchange value, as conceptual abstraction, is a manifestation of the conceptual subsumption of the manifold latent in the commodity. Hence, the commodity contains within it a manifold that is susceptible to historical determination; its nature will change based on historical circumstances, and different attributes will be externalised under different conditions.

Exchange value is a relation that changes with time and place and therefore appears as a semblance [*Schein*], which is relative and therefore constitutive of the thing the value represents. However, exchange value is determined by a common element between commodities and is thus a product of value-as-form. Since a given commodity will differ from another given commodity in its use value, the common element must be given by way of its quantity. Hence, the 'exchange relation of commodities is characterized precisely by its abstraction from [commodity] use-values'.[88] Once we have conceptually disregarded the use value of a commodity, what we are left with is only the property of the commodity as being a product of labour. According to Marx, commodities are:

> ... merely congealed quantities of homogenous human labour, i.e. of human labour-power expended without regard to the form of the expenditure. All these things now tell us is that human labour-power has been expended to produce them, human labour is accumulated in them. As crystals of this social substance, which is common to them all, they are values, commodity values.[89]

It is from this point that Marx determines that generalised social labour is the common factor in the exchange relation. This common factor, generalised social labour, is value (both value-as-content and value-as-form), and only becomes exchange value in its phenomenal form as a manifestation of value.[90] Marx goes on to explain the nature of value, independent of this form of appearance in its phenomenal form, to develop its measure as an exchange value. The measure of value is developed

---

87   Marx 1990, p. 125.
88   Marx 1990, p. 127.
89   Marx 1990, p. 128.
90   This reading of the term 'presentation' [*Darstellung*] is derived from Bellofiore's account in 'Marx After Hegel: Capital as Totality and the Centrality of Production' (2016, pp. 31–64).

... by the means of the quantity of the 'value-forming substance', the labour contained in the article. This quantity is measured by its duration, and the labour-time is itself measured on the particular scale of hours, days etc.[91]

Therefore, the measure of value is based on the quantity of the labour itself, deriving its meaning from labour time.

Value's magnitude is determined by the socially necessary labour time required for the production of a given commodity. Therefore, the objective character of exchange value is an expression of a homogenous social substance: abstract labour. The objective character of value is purely social in character, a character that is manifest only within a social relation between one commodity and another commodity, where the substance of value is abstract labour and the measure of its magnitude, exchange value, is labour time. What, then, is the form that 'stamps the value as exchange-value',[92] and therefore as a common value form? The missing link in this exposition so far is the money form.

As we have seen, money first appears in *Capital* Volume I as the form of appearance (appearing as a semblance) of abstract labour. As Fred Moseley has explained,

> ... because the abstract labour which Marx assumed to determine the value of commodities is not directly observable or recognizable as such, this abstract labour must acquire an objective 'form of appearance' which renders the values of all commodities observable and mutually comparable. This necessity of a common unified form of appearance of the abstract labour contained in commodities ultimately leads to the conclusion that this form of appearance must be money. Money is not an inessential illustration for labour-time. Money is the necessary form of appearance of labour-times.[93]

If money is the necessary form of appearance of labour time, then money is a temporalised form constituted by the durational time of circulation. Further,

---

91  Marx 1990, p. 129.
92  This phrase is taken from *A Contribution to the Critique of Political Economy*, where Marx began to elaborate his ideas on money. These basic ideas were developed in *Capital* Volume I, based on conceptual changes tied to a renewed emphasis on language. As such, *A Contribution* is not here a primary conceptual resource for our understanding of money (de Brunhoff and Foley 2007, p. 131).
93  Moseley 2015, p. 3.

money as the form of appearance of labour time is the form of appearance of the value hidden in the commodity. The value relation appears as a social relation between one commodity and another commodity. This is based on their mutual hidden substance of value: labour. Yet, in order to establish commodity circulation, one commodity must be set apart from the function of being a relative form of value, as in the case of the commodity, in order to be an equivalent form of value. In this way, money takes on a form that is not reducible to a commodity form. Opposed yet inseparable, their respective value determinations are relative, as 'there exists neither value, nor magnitude of value anywhere except in its expression by means of the exchange relation'.[94]

To recapitulate, in their emergence, both money and commodity rely on the money commodity (gold) for their mutually constituting structural relations. However, money, while finding its genesis in the commodity, develops into a general form that is both a bearer of the universal measure of value and the form under which value can circulate from one form to another. What de Brunhoff emphasises in her monetary theory is that money becomes a general form, set apart from the commodity form.[95] Marx construes money as a necessary component of the structural composition of the value form, as derived from the measure of labour time,[96] and the constitution of the commodity based on its fetish character as both use value and exchange-value. 'Commodity' and 'money' are mutually constituting within the logical self-development of value, where money plays a necessary role as both a general equivalent of value and the medium of the circulation of value. Money does this by mediating the change in forms of value through the application of measure that applies equivalences between things that are different. Therefore, money, as a commodity, allows value to change forms, from labour to commodity to capital.[97] Although money establishes its role based on its participation as a commodity that at once has a use value and an exchange value – measuring exchange by implementing a standard of price based on the measure of its own equivalence to the price of a commodity – money's formal character is not this simple, because money also circulates as credit.

---

[94] Marx 1990, p. 153.
[95] See de Brunhoff 2015.
[96] While value as content is measured by labour time, value as form is measured by money, producing a monetary expression of labour time. The measurement occurs in exchange; this is ideally anticipated within the production process (Bellofiore 2005, pp. 124–39).
[97] While currencies have a price in their exchangeability between one another, money qua money has no price, another attribute differentiating it from the commodity form.

Money, a form of value without a price of its own, functions to determine the price of commodities due to its nature as inversely proportional to all other commodities. As a form that functions to present value, money is the necessary mediator and the mutually constitutive form that facilitates the possibility for the commodity to come into being as a commodity. Money provides the commodity with its measure and its formal exchangeability only because it is itself a commodity in its conceptualised material form (gold). This universal equivalent form, gold as commodity, becomes money through social custom. It became universal, and therefore transformed into the money form, only as it gained a monopoly as the presentation of value as the content (*Darstellung*) of commodities.[98] As de Brunhoff shows, 'gold is able to play the role of money in relation to other commodities because it has already played the role of commodity in relation to them'.[99] Therefore, the historical reason for metal to function as money is logically subordinate to its theoretical reason for operating as at once a commodity and not a commodity.[100] While money develops as a money commodity, it must also remain something different from all other commodities, something set apart. The general equivalent must at once remain a commodity to the extent that it 'acts as if' it is a commodity, relationally, while differing from all other commodities since, although it has value, it has no price. Much like Fichte's 'I = I plus non-I', the general equivalent contains its own opposite. De Brunhoff emphasises that 'without this, every commodity would be money and all money a simple commodity, so that there would be neither money nor commodity production in which private exchange presupposed private production'.[101] Due to this role money plays in the development and intelligibility of value – where money applies a measure of value to commodities in which a price is derived, and where money has no price of its own – we find that money is at once the most basic and abstract form of value.

De Brunhoff's work on money reflects the way in which the commodity does not exist without the mutual constitution of money as the form taken by abstract labour. Therefore, for de Brunhoff, as in Marx, money is the form that facilitates the capitalist mode of production due to its role in the extraction of surplus value from living labour. The production of surplus value, or 'the increment of money that emerges at the end of the circuit of capital',[102] is the purpose of capitalist production. However, de Brunhoff at the same time

---

98  Marx 1990, p. 163.
99  de Brunhoff 2015, p. 23.
100 Ibid.
101 Ibid.
102 Moseley 2015, p. 9.

emphasises the need for an internally differentiated understanding of money's role as not being a strictly capitalist form (although being immanently capitalist).

What follows from de Brunhoff's position is that a significant aspect of the money form is marginalised in a purely value-form reading (that focuses on money's fetish character). Value-form readings can thus overlook the implication that, due to the structure of the value form, money is a mechanism at the heart of capital relations with social, economic and political consequences that *exceed* money as a mere phenomenal appearance of value, especially in its more developed forms within the credit system. The emphasis on money as a phenomenal appearance of value, and the resulting centrality of commodity exchange, forces the mediations that structure the analysis of value form to the periphery. These mediations include the social, economic and political implications of historically specific forms of capital accumulation that underpin the role of the state, wage relations, social reproduction, technological change and modes of labour (paid and unpaid).

### 2.2.4   *Money's Externality*

De Brunhoff was heavily influenced by *Primo Maggio*'s workgroup on money's focus on the management of money by the bank and the state, a focus she had neglected in her book, *Marx on Money*. Her later book, *The State, Capital and Economic Policy* – originally published in French in 1976 – was informed by her engagement with the workgroup, especially its conviction that 'the crisis of the international monetary system cannot be understood without a strong institutional and political component being incorporated into Marx's analysis of objective laws'.[103] As a result, the engagement between the two sides – *Primo Maggio* and de Brunhoff – provided an unexpected, underlying link between these two seemingly discrete traditions of Western Marxism, bestowing helpful tools for understanding still largely unresolved questions in our current context. Importantly, what both the *Primo Maggio* group and de Brunhoff understood was that, in order to make sense of the intensified role of the credit system in the context of the suspension of the dollars convertibility in 1971, and the corresponding increased circulation of financialised forms of money, analysis needs to consider the ways in which production and consumption are financed and determined on an institutional basis that is not strictly capitalist.[104] De

---

103   Lucarelli 2013, p. 9.
104   The Italian 'theory of monetary circuit' (TMC) also reflects this contextual and theoretical convergence, where a theory of value is developed without commodity money. Figures associated with this reading include Augusto Graziani and Riccardo Bellofiore. In

Brunhoff's conceptualisation of 'money as money' reflects this dynamic, in its very form, because money acts as money qua money, and not as commodity or capital, during certain moments within the process of circulation. Yet, de Brunhoff's standpoint maintains that the logic of capital is the overarching foundation of the money form: a standpoint derived from her attention to Marx's presentation of money across all three volumes of *Capital*.

De Brunhoff's monetary theory, therefore, provides the conceptual tools to analyse not only the role of banking capital and interest-bearing capital from the point of view of the money form, but to address how these forms and institutions are implicated in both the realm of production and the reproduction of capitalist societies. These forms and institutions base themselves on social relationships that are not strictly determined by the fetish character of the value form. Thus, they oblige a mode of analysis that understands money as a distinct form that is both separate from and immanent to capital.

This project requires further attention to the role of non-capitalist institutions[105] necessary for the reproduction of both social life (through the wage relation and de facto the reproduction of life qua life of members of society) and the reproduction of money as a general equivalent. These two sides – social life and money – are effectively reproduced through the movement of money's circulation. Non-capitalist institutions intervene on the analysis of

---

solidarity with the work of Marcello Messori of the *Primo Maggio* workgroup on money, Bellofiore, who also worked with the group, developed a reinterpretation of capital that retains the capital-labour relation while also claiming that money is not a commodity. This is developed through an interpretation of socialisation [*vergesellschaftung*] in Marx, where Bellofiore claims that there are three different concepts of socialisation formed in relation to capital's valorisation: (1) *ex post*: the socialisation that occurs on the commodity market at the point of the final exchange or final monetary validation; (2) the 'immediate socialisation' occurring with the immediate production process; and (3) *ex ante*: the monetary validation that initially takes place through the banking system at the point of sale and purchase of labour power. The third form of socialisation, added by Bellofiore, is based on what he calls 'anti-validation' and functions to integrate the role of the bank, as financing production, into the dynamics of valorisation. To do this, bank financing has to be understood not only as based on the socialisation of abstract labour, but also on the basis of a concept of socialisation different to those of *ex post* and 'immediate socialisation'. This third form of socialisation is based on a monetary theory of value without money as a commodity (see Bellofiore 2005).

105  The category of non-capitalist is used here in a way that is antithetical to the predominant view that an institution is capitalist because it is functional in reproducing capitalist social relations. This book instead claims that there are necessarily non-capitalist variables internal to capitalism. Non-capitalist variables are a structural aspect of capitalism and therefore necessary to theorise if we are to understand the range of modes of subjection, exploitation and domination within capital's social relations.

credit money through the dynamics of the value form. Because credit money is advanced before final realisation of capital, credit money is not valorised value and therefore is not dictated by the autonomy of the value form. Outside of capital but immanent to it, according to de Brunhoff, 'non-capitalist' institutions are structurally involved at the level of credit and finance (such as the state, central bank and other financial institutions). They are so on the very basis of Marx's formal account of value. It is through commitment to understanding 'money as money' that we can develop a robust account of social form and the internal complexity of subjection to capitalist social relations that are not determined by the fetish character alone.

De Brunhoff's Marxist monetary theory enables a renewed interpretation of capital, especially in its financialised form. Hence, I have argued for the need to ground analyses of capital in a concept of 'money as money'. In the context of heavily financialised markets that structure social life today, this mode of inquiry is indispensable as more and more subjection becomes determined by the circulation of credit money – and not through money that is formally valorised. Therefore, although it is necessary to understand these non-capitalist forms of subjection as rooted in the fetish character of the capital relation – as is the case in de Brunhoff's account of money as immanent externality – comprehending the ways in which social life is subject to and reproduced within capitalist social relations now requires greater attention to the interplay between capitalist and non-capitalist forms and institutions. What constitutes such forms and institutions is theorised in the rest of this book, departing from a distinction between impersonal and interpersonal forms of domination. This is addressed in Chapter 3, 'Fetish Character'.

CHAPTER 3

# Fetish Character

### Introduction

A correct interpretation of the logic of capital's reproduction requires comprehension of the different value forms' fetish characters. As this chapter will contend, the process of reproduction sustains the fetish character of capital's relation, and vice versa.[1] Therefore, interpreting the logic of capital's reproduction with attention to the fetish character of the different value forms permits a full understanding of money's fetish role, where money is a medium of reproduction of the general fetish character of society as a whole. Society's general fetish character is the mystification of the bourgeois relations of production, which are sustained by the respective particularities of the fetish character of each value form.

The social formation of fetishised relations is unique to capitalism because, there, the economy is directed towards the accumulation of value and not the circulation of products for use. This chapter analyses how the asymmetrical inversion of the relation between persons and things – characterising the fetish relation – functions within the different value forms. Differences between the fetish characters and mystification within the different value forms confer meaning to differentiated forms of subjection within capital's reproduction. Here, 'fetishism' denotes the specificity of the capitalist relation and therefore the combinatory function of the money fetish, the commodity fetish and the capital fetish; it is what makes distinctly capitalist relations impersonal abstract social relations – in which things represent social relations and are personified, and persons are reified and thing-like.[2] The resulting formation of social relations is an economic relation between people indirectly mediated by the commodity form. This dynamic differs from extra-economic interpersonal relations in which individuals interact with one another, albeit as abstract juridical persons. The dominance of commodity fetishism, as the social form of general social relations, however, presupposes other categories and relations, including those of juridical persons.

---

1 I.I. Rubin develops a similar claim in his book *Essays on Marx's Theory of Value* (2016, pp. 22–5).
2 See 4 of Chapter 1, Volume I, 'The Fetishism of the Commodity and Its Secret' (Marx 1990, p. 163).

In capitalist social relations, the state endows each person with juridical equality while at the same time protecting private property. Because private property is allocated along class lines, the state's neutral claim to provision of judicial equality to each person is in practice only an equality between persons within a social class. While these persons might not relate across classes with equality, within the social relations which require juridical mystifications, they can still relate to one another interpersonally (from kinship relations to extra-economic forms of violence). Interpersonal – and, in their isolation, non-capitalist – relations persist within the framework of bourgeois society's mystification as they are produced and reproduced by capital's fetishised, impersonal forms.

Capital's social relations include both capitalist and non-capitalist relations as well as personal and impersonal relations. The purpose of this chapter is to decisively theorise this distinction, mobilising these definitions for an improved understanding of the forms of relations internal to capital's reproduction process. To do so, this chapter, in the opening section 3.1, 'The Presupposition of Reification and The Money Form', first explains the interpretation of the fetish character through a discussion of the impersonal social relation of reification. This is followed by a Hegelian value-theoretical interpretation thereof in the subsequent subsections 3.1.1, 'The Hegelian Movement of Value as Essence'; 3.1.2, 'Value as Essence in Labour Time'; and 3.1.3, 'The Driving Force of the Process'. These subsections establish the logical character of reification as the movement of value between forms as their movement is analogous to Hegel's exposition of dialectical logic. Therefore, a focus on Hegelian philosophy provides a much more substantial understanding of the dialectical logic of the value forms themselves that in turn enables a sharpened analysis of their social consequences. The character of reification is then analysed in distinction from generalised fetishism and mystification as explained by Marx in his writings on 'the trinity formula'. In the subsections 3.1.4, 'The Fetish Character of Money', and 3.1.5, 'The Automatic Fetish of Interest-Bearing Capital', the chapter proceeds to discuss the specificity of the role of money, claiming that Marx's theory of commodity fetishism is a monetary theory of value in which interest-bearing capital retains a form of fetish that operates at a distinct level of abstraction from the money form more generally. Finally, in the second section, 3.2, 'Personal and Impersonal Forms of Domination', the chapter's initial moves are used to conceptualise the distinction between personal and impersonal social relations by deploying Evgeny Pashukanis's work on the legal form in *Capital*, and to consider the place of non-capitalist social relations within the logic of capital and its fetish.

## 3.1     The Presupposition of Reification and The Money Form

The fetish character of capital's abstract forms is the consequence of reified social relations that are objectified in things (commodity, money and capital). Within these relations, value moves between its fetishised forms as an abstract subject, acting independently of the will and actions of those engaged in exchange. Social life thus takes place through the movement of things, which controls the movement of human actors in the process.[3] Therefore, value is an expression of a social relation that is not controlled by individuals but by the relational form, which itself appears as things, which themselves are personified. The capitalist, for instance, is the personification of capital. The abstraction of capital renders the capitalist's status as a person thing-like because they are determined by the function of the abstract social form. These relations of domination, engendered by value, are not interpersonal but impersonal, because individuals only relate to one another indirectly through the mediation of things.

The value of commodities is an expression of a social relation where the relations of domination are objective [*sachliche*] and not interpersonal. This objectivity is produced by an inversion of persons and things that gives 'subjectivity' to the movement of value (between the different forms) and reified objectivity to individuals. While Marx uses 'fetishism' to explain the character of the social process as a whole,[4] the inversion behind the appearance of the value forms are the only instances of the fetish itself and, within the process of valorisation of capital, are only formally secured as having a fetish character at the point of exchange.[5] The valorisation of value occurs at the point of exchange and, as such, retroactively secures the social form of capital and its fetish. Therefore, it is the exchange abstraction that posits value as a reified essence.[6] As forms possessing a fetish character, value forms are reproduced through the self-movement of capital, while capital's self-movement subjects individuals to its form by its determination of everyday practices.

Within capitalist social relations, human labour is measured to create abstract labour (the substance of value). In doing so, the substance of human

---

3   Marx 1990, pp. 169–70.
4   Best 2015, p. 123.
5   This argument draws on the works of Rubin and the subsequent readings of the Neue Marxlekture, in particular Hans-Georg Backhaus's in his essay 'On the Dialectics of the Value-Form' (1980), and subsequent generations that have taken on value analytical frameworks, such as Michael Heinrich, Christopher J. Arthur, and Elena Lange.
6   Arthur 2004, p. 95.

labour is transformed into value as the essence of capital. The fetish character of capital is the character of the social relation of capital produced through the person-thing inversion, which creates value as essence. Ultimately, it is the relationship between measure and essence that makes such an inversion possible. For labour to be measurable, it must be abstract and homogenous, or general, requiring productive labour to be measured in relation to total social capital. Counterintuitively, it is the individualisation of commodity producers (including workers whose labour is the commodity they bring to the market, where there is an absence of direct social activity) that retains the indirect connection between individual labour and social labour. The individualised products of labour, when exchanged, come to abstractly represent the labour of society as a specific measure of labour in the abstract. The interconnection of social labour and concrete labour is established through the exchange of individual products.[7]

Capitalist exchange is only possible when labour is abstracted from concrete labour (which is particular) into abstract and homogenous labour (which is a general product of social production relations as a whole). The socialised nature of labour, internal to value, permits the 'exchangeability' of concretely different things. The heterogeneity of the different value forms acquires a homogenous essence based on this socially general labour, 'abstract labour', which is the collective abstraction of labour as a quantity based on measure, not its particularity as practice. The money form determines the value content of abstract labour and therefore facilitates the reification of concrete labour into a substance that is abstract and general. Exchangeability is possible because labour is abstracted from the concretely useful labour manifest in commodities and rendered homogenous. As Elena Lange correctly articulates, through exchanging two commodities that are entirely different, requiring distinct labouring activity for their production, 'the labour manifested in the commodity that is in the equivalent form becomes the incarnation or materialisation of value for the commodity that is in the relative form of value'.[8]

Within developed capitalist relations, the equivalent form of value is money; it connects the other forms of value to one another. In its fetish character, the money form is a universal equivalent that both remains other to commodities and exists formally alongside them in a completely reified from: it is the abstraction of value representing itself. Its abstraction from concrete labour – to homogenous social labour – facilitates the exchangeability of all

---

7  Engels and Marx 1962, p. 461.
8  Lange 2014, p. 28.

other commodities. Money, therefore, appears without the mediation of its genesis in abstract labour.[9] This special role that the money form plays detaches all commodities from their concrete existence. Money purposes commodities for the passage of value within the circulation and valorisation of capital. In doing so, money facilitates the fetish character of social relations, rendering the exchange of money the social form of capital's reproduction.[10]

For money to facilitate the fetish character of social relations, reification must be presupposed. Money, the incarnation of abstract labour as value, is the precondition for the act of commodity exchange. And commodity exchange is only possible if money is already in existence as a reified abstract form. As a reified abstract form, money is not generated by exchange; rather, it is presupposed as a general equivalent. This general equivalent is what endows the whole process with a 'phantom-like objectivity' [*gespenstige Gegenstandlichkeit*].[11] The origin of this precondition is in the fetish character itself: across *Capital*, Marx clarifies that fetishism is an inversion arising from the structure of capitalist social relations and the corresponding practices of individuals in capitalist societies that together reproduce its form. Marx explains this by stating that the individuals have 'acted before thinking', reflecting how the social process is itself presupposed by reified abstractions. According to Marx, in engaging in an exchange an individual has,

> ... already acted before thinking. *The laws of commodity nature act upon the natural instinct of the commodity owners*. They can only relate their commodities to each other as values and therefore as commodities, if they place them in a polar relationship to another commodity as general equivalent. We concluded this from the analysis of the commodity. But only a social deed can turn one specific commodity into the general equivalent.[12]

The owners of commodities act according to a presupposed logic, whereby the commodity's 'laws of nature' determine the *natural instinct* of the owner of the commodity. Even before money is present, the 'social deed' confirms that the

---

9   As Lange has also rightly observed, 'in its completely developed and reified form ... labour becomes money. Money does not "leave a trace" of its own genesis – therein consists its magic' (2014, p. 28).
10  Rubin 2016, p. 9.
11  Lange 2014, p. 23.
12  See Lange, 2014. This passage is Lange's translation of Marx taken from *Capital* Volume 1 in the MEGA [emphasis is Lange's].

money commodity will be a general equivalent ahead of time. Thus, without knowing it, owners of commodities act in line with the theory that money is indispensable for exchange.

As this chapter has established, money form, as a general equivalent, is presupposed in the act of capitalist exchange itself; it exists with homogenous abstract labour as its substance.[13] The act of exchange anticipates the money form's spontaneous acquisition of its role. There can be no value accumulation without the money form providing the measure of value and its medium of circulation. Money is required as a general equivalent to connect the different forms in circulation. The money form is, in fact, the sole form in which the value of the commodity appears, since no value form exists independently of exchange. In this regard, the contradiction between use value and value is present even before money appears to determine it.[14]

### 3.1.1  The Hegelian Movement of Value as Essence

Understanding the inversion that constitutes the fetish character requires closer examination of the movement of value as essence. Doing so provides a better grasp of the change in quality of concrete labour to abstract labour. This is a moment of inversion underpinned by the contradiction between the life activity of the labourer, or the individual, and capitalist social form. Here, quantity (abstract labour) relates to quality (labour that has its basis in an ontology: the human individual); the two sides cannot be understood separately.

To analyse value as essence, it is instructive to delve into the Hegelian philosophy behind Marx's use of the relationship between quantity (abstract labour) and quality (concrete labour). Doing so deepens our understanding of Marx's analogous presentation of the value forms. It is instructive to use Stavros Tombazos's work on Hegel's theory of measure for this purpose. In his chapter 'The Hegelian Theory of Measure and Value as "Essence"', Tombazos clarifies the Hegelian logic beneath Marx's theory of value, as well as its reliance on the inversion of subject and object. Astutely, Tombazos grasps that this inversion occurs through the measure of labour time as that which creates value, which is essence.[15] The inversion of persons and things in Marx is here correctly under-

---

13  For commentary on this, see Samezō Kuruma, *Theory of the Value-Form & Theory of the Exchange Process* (1957, pp. 24–5). Kuruma is discussed by Lange in her article 'Failed Abstraction – The Problem of Uno Kōzō's Reading of Marx's Theory of the Value Form' (2014, p. 26).

14  This position is behind the interpretation of Marx given by Michael Heinrich in *An Introduction to the Three Volumes of Karl Marx's Capital* (2012).

15  Abstract labour is the substance of value, while value is the essence of the forms of value's appearance.

stood as a process of the becoming of value (which constitutes the essence of capital). As in the case of the becoming of essence in Hegel,[16] this process results in a qualitative reversal of form based on the input of measure. Tombazos stresses that the relationship between the quantitative and qualitative successions – which permit the creation and movement of value in Marx – is drawn from Hegel's presentation of measure and the becoming of essence in the first book of the *Science of Logic*.

In the section on 'measure' in the *Science of Logic*, Hegel discusses the link between 'quality' and 'quantity' (analogous to Marx's use thereof). Hegel outlines the progression of quantitative nature as a series of discontinuities that are qualitative. He refers to this as a 'nodal line of measure'.[17] What Hegel's 'nodal line' reveals is that qualitative discontinuities occur not through slight progression but 'qualitative leaps'. Drawing on the model of a chemical reaction, Hegel shows that chemical substances, while appearing as 'determinate things', are in fact incompletely isolated. Yet, it is their very isolated nature that is the presupposition for their combination with other chemical substances. Thus, Hegel observes that it is the quantity of their quality that becomes the requirement behind the saturation of one substance with another, leading to a marked change of quality upon the merging of chemical substances.[18] Illuminatingly, Tombazos discerns the strong similarity between this description and Marx's description of the quality of commodities. The qualitative nature of commodities is determined by the quantity that is required for their saturation or capacity to exist in relation to other commodities. As in Hegel's chemical reaction, where the quantity determines the qualitive nature of a given substance, the qualitative is determined by its quantitative aspect. A substance can only persist as a specific quantity. Following this logic, the social quality of a commodity is made up of quantitative value only through its relation to other commodities. The quantity that is required to determine the quality of a commodity (its saturation) are the commodities for which it will be exchanged.[19] This chemical analogy, derived from Hegel, is a helpful way to think about the appearance of the quantitative aspect of the commodity. This is because, as in the case of a chemical quantity that determines the chemical compound, the value abstraction is not merely an abstract representation but the quantitative appearance of value.

16  'Measure', the third section of the first book of Hegel's *The Science of Logic*, contains a chapter called 'The becoming of essence'.
17  Tombazos 2015, p. 43.
18  Hegel 2010, pp. 369–70.
19  Tombazos 2015, p. 44.

For Marx, commodities are equivalents in the same way that simple chemical elements combine proportionately to form 'chemical equivalents'.[20] Relations of exchange function as 'relational measures' of the implicit material substrate. If the quality of a commodity is its function as a regulatory principle, or general equivalent, its material substrate is the physiological aspect of abstract labour: the consumption of labouring individuals made up of muscles, nerves and brains.[21] In 'The Becoming of Essence' in *The Science of Logic*, each combination of measures is understood as a qualitative state of the substrate (such as in the case of the substrate of water, which might take the state of liquid, solid or gas). Therefore, while there may be a qualitative change, it is a change of *state* only. The *subject* of this transition remains the same.[22] The substrate is therefore a 'regulative principle' that retains self-identity regardless of how it expresses itself qualitatively. The substrate effectively has an indifference to its determinacy, which is expressed quantitatively and qualitatively.[23] These states differ in quantity and quality. Therefore, 'the substrate itself, as an indifference, is likewise in itself the unity of both qualities'.[24] For Hegel, this unity, resulting from the indifference of the substrate, is a negatively posited contradiction. Dialectically, this negative positing is sublated into a self-subsistent 'being', an 'immanently negative and absolute unity which is called essence'.[25] Thus, essence is a self-determined, negative absolute unity.[26]

This interlude into Hegelian philosophy is helpful for clarifying the nature of value as essence in Marx's exposition, for Hegel exemplifies how essence neither exists at the level of being nor at the level of immediate objectivity. As such, objects contain a property that does not belong to their materiality. This is an aspect of Marx's theory of value that is not only difficult to understand but has few equivalents in social thought. Marx's exposition of value as essence therefore becomes more easily interpreted when made analogous to the chemical composition in Hegel's construction. It is this process of becoming-essence that explains the movement of value, whereby value exists as the essence of capital. Capital's essence, although real, does not exist at the level of being or immediate objectivity, but at the level of quantitative abstraction.

---

20  Marx 1970, p. 34.
21  Tombazos 2015, p. 46. The physiological aspect of the human individual will be developed in the final chapter, 'Marx's Social Theory of Reproduction', in section 5.4, 'Concrete Reproduction of Human Life and Nature'.
22  Hegel 2010, p. 373.
23  Tombazos 2015, p. 46.
24  Hegel 2010, p. 376.
25  Hegel 2010, p. 379.
26  Tombazos 2015, p. 46.

### 3.1.2  Value as Essence in Labour Time

Because capitalist social relations posit their own presuppositions, the inversion behind the fetish character of capital is, in the first place, determined by its product: social capital or the total capital accumulated by the social form. Arising through the mediation of the commodity, social capital is the product of the fetish inversion, forming the link between labour time and value. Value links the labour time of production with the social capital used to purchase commodities produced by labour time. The commodity, a necessary mediation as value (essence), does not exist as a being or in objectivity; rather, as in the case of essence, value is negatively posited in indifference to its objectivity (state) and therefore resides in different forms, retaining its own nature as self-subsistent. Value is 'socially necessary abstract/general labour time'[27] that, through its measurement (by the money form), forges the relationship between a commodity and the consumer who purchases it to meet a social need. Value as a self-subsistent being is a form which finds itself to be a state and therefore a reflection of the sphere of being. What is retained when value is reflected in other value forms (states) is abstract labour, or social labour time. Abstract labour, therefore, forms the substance of value.

The asymmetrical relationship between the value of labour and the value of the commodity co-produce capital's fetish. Because the wage only pays for a portion of the value produced by labour time, an asymmetry arises which transfers the subjectivity of the labourer, by way of the expenditure of labour power (measured as abstract labour), to the objective movement of value forms. If this relationship was equal, value would not accumulate more value and would not reproduce itself. The creation of more value occurs through the abstraction of surplus value, which is the part of labour time that is not paid. This asymmetry inverts the driving force behind the social relation from the human individual to the reproduction of value in its realisation. Value is a social relation that is 'autonomous and dominant' because the value of labour is not made equal to the value of the commodity.[28] Therefore, as we can see, much like how the movement of essence in Hegel entails a qualitative change though a quantitative relation, measure is a crucial factor behind the 'qualitative leap' made when value moves (as abstract labour) from labour time to the commodity.

---

27   Tombazos 2015, p. 53.
28   Tombazos 2015, p. 53.

### 3.1.3 *The Driving Force of the Process*

While the qualitative leap from labour time to the commodity produces the reification of persons and the personification of things, value – as essence – lies behind the transition: by retaining its self-identity, value as essence makes this metamorphosis possible. Without the continuity of essence, there would be no inversion. This inversion – which is established by the lack of equilibrium between value produced and value accounted for in labour time by wages – gives subjectivity to the objective process of value's accumulation. Without this, value would not be accumulated; instead, it would be expended within the process. As such, the production of commodities is directed towards the accumulation of value with indifference to the particularities of consumption. Accordingly, accumulation and the reproduction of value are the driving force of the process, not commodity production supporting human subsistence. Human life becomes an object instrumentalised for the accumulation of more value and, thus, for capital. For Marx, abstract labour is not simply socially necessary labour time embedded in commodities, but is also an 'autonomous subject' that is produced in the movement of the value form as it reproduces itself. This movement of self-reproduction is a movement of things without the control of individuals. It is 'a movement made by things, and these things, far from being under [an individual's] control, in fact [controls] them'.[29] The social relations of production not only take the form of things but can only be expressed through things.

Although abstract, the inversion constituting the fetish is a real social relation. Importantly, while the fetish character produces a false consciousness – which veils social relations with things – it is not a false reality. Social relations *do* become autonomous things and dominate individuals, subjecting them to their movement: this characterises capitalism as a mode of production.[30] 'False consciousness' is a necessary component of the atomisation of this social relation. Individuals necessarily represent their social relations in a false manner (individuals *are* thingly, as character masks or personifications of economic categories) because this very false representation is a part of the social relation itself.[31] As a result, those within the capitalist mode of production live in a world that is 'enchanted' with the 'personification of things'. Capital, the thing, appears as a person: the capitalist as bearer of capital's form. The subjects of

---

29   Marx 1990, pp. 167–68.
30   Tombazos 2015, p. 24.
31   Ibid.

this social process are the fetishised commodity forms: commodity, money and capital – and not 'persons' endowed with 'freedom'.[32]

The subjection of people to the movement of value between these reified forms does not occur merely due to false consciousness. Rather, consciousness is produced by social practice, where value's accumulation is acted out. Social practice is determined by the objective means of accumulating more value, and so production takes on a life of its own. Personified things contain a material force that gives individuals such a powerful, objective duty to perform that failure to do so would entail deep suffering and 'ruin'.[33] Thus, it is crucial to comprehend the subject-object inversion. The subject-object inversion is a key moment in the development of 'capital as subject' and the reproduction of the 'life' of capital as self-moving. This is a real process that requires false representations to appear before those who are subject thereto.

### 3.1.4    The Fetish Character of Money

The interpretation of *Capital* developed in this book is interested in the different modes of subjection to capitalist social relations understood from the standpoint of an expanded perspective of the reproduction of capital. Because the money form is the medium of circulation connecting different elements to one another in exchange, it is necessary to pay close attention to the role of the money form and its specific fetish character.

Money is the only form in which value exists independently of exchange. This is because money is the only form in which the value of the commodity appears: although other forms might independently have a price, no other form can manifest value without being put in relation to the money form through exchange. What this means is that within the movement of value forms, or circulation, there is no longer any differentiation between labour that is expended privately and abstract social labour. The lack of differentiation between privately expended and socially recognised labour elucidates the role of abstract labour as a homogenous substance, which is not a natural attribute of labour and only exists when represented in commodities. Value is the negative positing of the contradiction between concrete and abstract labour, while the fetish character of value forms is determined by monetary circulation.

---

32  Marx's definition of the person is explained in Chapter 2 of *Capital* Volume I, 'The Process of Exchange'; there, he draws on Hegel's legal definition of the person as a person as endowed with freedom, or 'right'. A discussion of what constitutes a person will be discussed in more detail in the final section of this chapter, 'Personal and Impersonal Relations'.

33  Heinrich 2004, p. 185.

The monetary theory of value differentiates Marx's writings from classical political economy. The difference is exemplified in his form analysis, which centres money as the only value form in which the value of the commodity appears; money is the necessary expression of value's objectivity.[34] Consequently, Marx's interpretation of money permits an understanding of the movement of value from the point of view of capitalist social relations, whereby human subjection is differentiated within the process of valorisation. Rather than interpreting the fetish character of value forms as merely a form of false consciousness or instance of human alienation, attention to money allows analysis to grasp that the fetish character is a product of social relations as a whole. As a multi-faceted conceptualisation of objective social domination, Marx's theory of fetishism shows us that commodities both possess 'value-objectivity' and are 'spectral' as a result of the practical organisation of capital's social relation.

Marx's theory of fetishism illustrates how the value forms of commodity, money and capital possess a particular character. These forms have a 'fetish'[35] character because they are social relations that appear to possess a thing-like objectivity. This thing, the commodity, money or capital, is a real objectification of a social relations. Therefore, it *is* a fetish; its fetish character is not 'ideological'. Within capital's social relations – which are fetishised social relations insofar as they are relations determined by the circulation of value forms – there are further mystifications, where value appears not as it really is (as is the case of the fetish). Marx argues that mystifications (not fetishes) arise when value is thought to exist where it formally does not. For example, wage is a mystification because it is a payment for the value of labour; and, while labour is represented by value, labour itself has no value.

Fetishism can occur only by way of the circulation of mystifications, which is identified by Marx as the 'trinity formula'. Therein, wage (labour), rent (land/nature) and interest (capital) are taken together to reflect distinct forms of income, each representing a different class relation. The trinity formula is a set of mystifications that are necessary conditions (and not theories or ideologies) for the fetish character. These mystifications are both practical presuppositions of capital's production process and its result (or creation).[36] In Chapter 48 of Volume III, 'The Trinity Formula', Marx states,

---

34  This theoretical standpoint is referred to as the 'monetary theory of value', which originated with the work of Backhaus and has been further developed by Heinrich and others.
35  For a genealogy of Marx's use of the term fetish, see Chapter 1, 'Commodity: Fetish and Hieroglyph' in Peter Osborne's *How to Read Marx* (2005, pp. 17–21).
36  Marx makes this claim in *Capital* Volume III on p. 957 (1991).

> The capitalist production process proceeds under specific material conditions, which are however also the bearers of specific social relations which the individuals enter into the process of reproducing their life. Those conditions, like these social relations [the trinity formula], are on the one hand presuppositions of the capitalist production process, on the other its results and creations; they are both produced by it and reproduced by it.[37]

The material basis of production resides in its ontological preconditions – human life, nature and the money form – which in their historical specificity are retrospectively posited by capital. The ontological preconditions of capital are grounded in life and therefore also in the reproductive processes of the life of these preconditions (be this the 'life of capital' as self-moving value or nature and human life).

Attention to the monetary theory of value within Marx's writings enables analysis to conceive the reach of the fetish character of capital's social relations. The fetish character, as a general mystification of bourgeois relations, binds all three volumes of *Capital*: all members of society are subject to the fetishism of social relations that emerges objectively and structures individuals' actions and perceptions. Due to the practical and not merely ideological nature of fetishism, none – neither capitalist nor workers – are subjected in such a manner that they evade its effects.[38] Here, the monetary theory of value is employed to interpret the role of the reproduction of human life in relation to the 'life of capital'. From this perspective, individuals are subjected to reified social relations as bearers of the capital relation within capital's circulation (its movement towards valorisation as a whole). Individuals are not merely subjected to capital's relations as a variable for the reproduction of productive labour. As such, the monetary theory of value harbours distinct political implications. Whilst traditional, production-oriented Marxisms have often overlooked the connection between the different movements of capital by developing a radical interpretation of form, i.e. the labour theory of value, the falsely determined opposition between speculative financial markets and capital's production process can be overcome by the monetary theory of value. Thus, the interconnection between finance and production can be brought to the fore through a monetary reading.

---

37  Ibid.
38  Heinrich 2004, p. 185.

The monetary theory of value does not grant a privileged position to the forms of subjection manifest in productive labour – or to that of the nature of subjection to financial operations; rather, it views productive labour as one form of subjection to the capital relation, co-existing with others in circulation even if it remains ontologically central to the creation of value. This permits a better understanding of the role of social reproduction – or the reproduction of human life – in that it does not restrict analyses' focus to the reproduction of productive labour. As such, it allows one to better understand the forms of subjection determining the unpaid social activity that upholds human life. Likewise, one can begin to conceive subjection to the growing amounts of fictitious capital in circulation.

Because the value forms reflect different moments in the process of valorisation, they produce different kinds of fetish: the commodity fetish, the capital fetish and the money fetish. The specificity of the money fetish, as a medium of reproduction ('money as money' and not 'money as capital' which takes on the fetish character of capital), begets its ability to appear as fictitious capital. As fictitious capital, the money fetish appears as value representing value, and it therefore appears as detached from its genesis in the production process as unmediated value or as already valorised before mediation in production. The money form appears immediately as 'capital in its finished form, the unity of the production and the circulation process'.[39] Marx explains as much in Volume III. There, he states that capital appears 'immediately' in the form of interest-bearing capital. By 'immediately', he means that capital appears unmediated by production and circulation. The appearance of capital as interest-bearing capital is therefore both 'mysterious' and 'self-creating' as a source of interest: it is a source of interest 'of its own increase'. Money produces more money (M-M$^1$), i.e. self-valorising value without the process of production as mediation. Although commercial capital (M-C-M$^1$), too, eludes production, the commodity at least remains present as a mediation (and the said commodity's conditions of production persist too), even if this takes place in circulation only. M-M$^1$ presents itself clearly as a 'social relation' and not a product of things. Marx explains:

> we have M-M$^1$, money that produces more money, self-valorizing value, without the process that mediates the two extremes. In commercial capital, M-C-M$^1$, at least the general form of the capitalist movement is present, even though this takes place only in the circulation sphere, so

---

[39] Marx 1991, pp. 515.

that profit appears as merely profit upon alienation; but for all that, it presents itself as the product of a social *relation*, not the product of a mere *thing*.[40]

The thing-like character of money as interest-bearing capital is the fetish of interest-bearing capital. It is only made possible by the fetish character of money, which can appear as value representing value. As a thing, or 'money capital', capital itself 'becomes a commodity whose self-valorizing quality has a fixed price as expressed in the prevailing rate of interest'. It is when capital becomes a commodity that it

> ... obtains its pure fetish form, M-M¹ being the subject, a thing for sale ... [T]he surplus value it creates, here again in the form of money, appears to accrue to it as such. Like the growth of trees, so the generation of money (*tokos*) seems a property of capital in this form of money capital.[41]

This is only possible because money exists as an autonomous exchange value or as a medium of circulation. This makes money a form where use-value distinctions between commodities are rendered obsolete. As a result, the distinctions between what consists of the commodities and their productive conditions – in industrial capitals – are also obsolete.[42] It is this lack of distinction between capitals that gives the money form its independence.

### 3.1.5   *The Automatic Fetish of Interest-Bearing Capital*

The specificity of the fetish character of interest-bearing capital – as a distinct appearance of the money form generating rent – is not representative of value. It is described by Marx in *Capital* Volume III, Chapter 24, 'Interest-Bearing Capital as the Superficial Form of the Capital Relation'. The first sentence of the chapter states, 'in interest-bearing capital, the capital relationship reaches its most superficial and fetishized form'.[43] Marx later terms interest-bearing capital an 'automatic fetish', therein distinguishing its character from that of the money form. Marx makes this distinction because the form of interest-bearing capital appears to be devoid of the capital relationship itself: it seemingly creates more value without the mediation of the commodity's, and therefore capital's, social relation. Consequently, the form does not contain the capitalist

---

40   Marx 1991, p. 515.
41   Marx 1991, p. 517.
42   Marx 1991, p. 517.
43   Marx 1991, p. 515.

social relation, which remains internal to the money form in which interest-bearing capital it is rooted. Rather, the money form of interest-bearing capital is fictitious, producing value shorn of content. Marx explains that the overall production process of interest-bearing capital arises as a property of capital itself. It is up to the possessor of money as to whether they want to spend the money or lend it as capital.

The 'automatic fetish' of interest-bearing capital (M-M$^1$) is formal. As self-valorising value, it dispenses of any markers of its origin: the commodity does not appear in its mediation. Hence, Marx calls interest-bearing capital 'money breeding money'. Marx claims that money, in relating to itself, consummates a social relation that is pure form without its content:

> ... the result of the overall reproduction process [of interest-bearing capital] appears as a property devolving on a thing itself; it is up to the possessor of money, i.e. of commodities in their ever-exchangeable form, whether he wants to spend this money as money or hire it out as capital. In interest-bearing capital, therefore, this automatic fetish is elaborated into its pure form, self-valorising value, money breeding money, and in this form it no longer bears any marks of its origin. The social relation is consummated in the relationship of a thing, money to itself. Instead of the actual transformation of money into capital, we have here only the form of this devoid of content.[44]

The fetish character of interest-bearing capital is automatic because interest-bearing capital accumulates value without the mediation of the social relation retrospectively determined by the exchange relation. Consequently, interest-bearing capital cannot accumulate what constitutes the content of value: abstract labour. Interest-bearing capital is therefore 'a form in which the source of profit is no longer recognizable and in which the result of the capitalist production process – separate from the process itself – obtains an autonomous existence'.[45] While interest-bearing capital is posited by the capitalist social relation, it no longer contains this social relation due to its self-referentiality: it excludes the production process from its mode of reproduction. Marx explains:

> The fetish character of capital and the representation of this capital fetish is now complete. In M-M$^1$ we have the irrational form of capital, the

---

44  Marx 1991, p. 516.
45  Marx 1991, p. 517.

misrepresentation and objectification of the relations of production, in its highest power: the interest-bearing form, the simple form of capital, in which it is taken as logically anterior to its own reproduction process: the ability of money or a commodity to valorize its own value independent of reproduction – the capital mystification in the most flagrant form.[46]

Interest-bearing capital relies on the simple form of capital, and its corresponding social relations of reproduction, as logically 'anterior' to its own independent process. Simultaneously, however, it is 'independent of reproduction' of the capitalist relation. This is because it accumulates value not from the extraction of abstract labour as surplus value but through interest 'as the specific fruit of capital', or rent.

Because the fetish of interest-bearing capital is that it produces its own opposite, Marx saw credit operations as a dynamic of 'the abolition of the capitalist mode of production within the capitalist mode of production itself'.[47] Impersonal social relations become interpersonal forms of subjection and 'appear' as abstractions, which are produced due to the separation of these relations from the production process. Interest-bearing capital causes value to appear as if it is accumulated from capital and therefore from historically specific social relations. The appearance is not a progressive 'rational' process of capitalist development (as in the appearance of the commodity fetish). In the case of interest-bearing capital, the fetishised appearance of more capital accumulated from capital is a fetish only because it does not contain the real nature of capital: it does not contain value and therefore is fictitious. The redoubled fetish of interest-bearing capital can only be understood by uncovering the fetish character of money, where analysis departs from a consideration of abstract labour.

While the appearance of interest-bearing capital fetishistically hides the fictitious nature of its value, it does not hide the historically specific social relations behind its ability to create more fictitious value. This is because the lending of money, which is not an exchange, requires an interpersonal, legal contract. While the actual capital accumulated is fictitious, these social relations are not mediated by exchange. Therefore, while the individual, as a bearer of capital personified, lends their money, they engage in an interpersonal relation between respective bearers of the capital relation. In this manner, the

---

46   Marx 1991, p. 516.
47   Marx 1991, p. 569.

specificity of the fetish character of money resides in its ability to produce value not determined by abstract labour but an interpersonal legal form, in which money is lent as credit. Money lent as interest-bearing capital (to extract rent) commands interpersonal relations based on logically anterior impersonal relational forms.[48] M-M$^1$, money creating more money, is the reduction of the formula M-C-M$^1$ to its two extremes. M-M$^1$, according to Marx, reduces the general formula for capital into an 'abbreviation'[49] that lacks content specific to capital's social form. In this sense, interest-bearing capital remains other to the accumulation of capital, as an abbreviation of the process, devoid of its content.

The reduction of the general formula of capital into an abbreviation is the result of the way in which the contradiction of capitalist and non-capitalist variables, determining reproduction, express themselves in their fetish character. Credit money does not exist as a form of value as a result of accumulated past abstract labour that puts the subject-object inversion of the fetish character into motion; rather, this form of value is produced through interpersonal forms of domination that promise future production of value as substance. Here, the forging of a contract takes the place of abstract labour in informing its substance (as the legitimising factor).[50] It is therefore instructive to consider credit, banking and fictitious capital in greater detail.

Banks, as institutions providing finance, exist through circular banking operations. By virtue of their circular nature, the banking system both maintains itself and reproduces itself. Purely financial circulation develops by giving the credit system a fictitious character. The fictitious character of the credit system at the same time 'preserves' the financial system that revolves around itself. This dynamic is cogently articulated by de Brunhoff when she claims,

> ... the function of the banks in financing rests on the circular character of banking operations, by virtue of which the banking system maintains and reproduces itself. At the same time a purely financial circulation develops; while it gives credit system a 'fictitious' character, it also preserves it as a financial system. Although the circuits of financing remain in the last analysis dependent on the needs of productive capitalists, they can endlessly revolve confusedly about themselves, independent of the cir-

---

48   Money lent as interest-bearing capital still requires a valorised referent.
49   Marx 1991, p. 515.
50   Therefore, Marx called interest-bearing capital fictitious. Interest-bearing capital lacks value's defining substance on account of its independence from production.

culation of capital ... [T]hough adopted to the needs of capitalism, credit is never really contemporaneous with capital.[51]

Fictitious capital is doubly determined by the value form that conditions it and the private economic relationships internal to contracts that account for anticipation of future modes of life required for repayment: future labour and social reproduction, both present and future, need to reproduce the indebted collective and individual members of capitalist society as able to pay in the future.

The forms of subjection corresponding to fictitious capital place distinct stress on the reproduction of human life as a component of the reproduction of capitalist social relations. Here, human life is not reproduced strictly through the mediation of the exchange relationship generated by the circulation of capital, but also in subjection to un-valorised value: an interpersonal form of social relations forged onto pre-existing impersonal relations, i.e. a redoubled fetish. This form of subjection occurs through a contract that requires the reproduction of human life to suit the purpose of this interpersonal relationship, sustaining financial circuits that evade the production process. Therefore, the fetish of the money form and its mode of domination as a medium of reproduction – where the circulation of credit plays an essential role funding means of production – is not simply that of impersonal relations existing through capital's abstraction. The fetish of the money form also contains interpersonal relations of domination. Marx emphasises this when he states:

> ... the credit that the reproductive capitalists give one another, and that the public give them, he makes into his own source of private enrichment. The final illusion of the capitalist system, that capital is the offspring of a person's own work and savings, is thereby demolished.[52]

Here, the logic of capital creates forms of subjection that are not developed through productive labour. This is possible, according to the logic of capital, only because money, as the medium of reproduction with the capacity to be interest bearing, is also other to the logic of capital. The fetish function – the 'automatic fetish' – of interest-bearing capital produces interpersonal forms of domination in addition to the category of the 'person' created by capital's fetish more generally.

---

51   de Brunhoff 2015, p. 98.
52   Marx 1991, p. 640.

## 3.2 Personal and Impersonal Forms of Domination

In one of the few systematic accounts of legal forms within Marx's critique, *Law & Marxism*, Pashukanis correctly states that 'legal fetishism compliments commodity fetishism'.[53] This is the case because the 'person' of the exchange relation is an abstract product of the relation as a property owner and therefore as a legal person. In defining the category of the person and subsequent forms of domination of the individual that depend on the category, the analysis offered in this section builds on Pashukanis's statement to claim that legal mystifications are interpersonal forms of domination based on the interrelations of persons with abstract rights to property (abstractly equal and with free will). Legal relations complement commodity fetishism as a necessary internal process of subject production, engendering the reification of 'persons'. Interpersonal forms of domination are understood here as social forms, which, while lacking the direct mediation of the commodity form, are socially necessary forms of appearance. Interpersonal domination, and the corresponding category of the person, are internally necessary to capital's social form.

### 3.2.1 *Marx's Use of the Category Person*

Marx employs the category of the 'person' in his articulation of commodity fetishism; the 'person' appears as a creation thereof. In capital's social relations, therefore, individuals relate to one another through interpersonal forms that rely on the category of the person, even if their relations are subsequently mediated by 'things'. The term individual is deployed consistently in this context to contrast a transhistorical entity with the person produced by capital's social relations. Throughout his critique, however, Marx employs the category of the person unsystematically. The category of the person is not only used by Marx to indicate the place of the human individual within the social relations of commodity fetishism, where the person is founded upon 'abstract right' to own property (necessary for commodity exchange); it is also used to indicate a subject position both before and after the alienation imposed by capital's social form. In doing so, Marx evokes the person as a subject position free from the alienation endured within capital's abstract form.[54] Marx's use of the person in the sense of the subject position before capitalist alienation can be located in

---

53  Pashukanis 1989, p. 117.
54  In the *Grundrisse*, Marx explains the following, indicating that there are interpersonal forms of domination and impersonal forms of domination. The former are carried over from pre-capitalist relations and therefore would warrant a different concept of the person than that produced by capitalist social relations. By contrast, a free individuality (not

the following examples from *Capital* Volume I. Here, the person is not referred to as impersonal inversion between subject and object – a 'personification' or a 'character mask' (economic or legal) – but as the bearer – or 'person' – themselves:[55]

> Let us now transport ourselves from Robinson's island, bathed in light, to medieval Europe, shrouded in darkness. Here, instead of the independent man, we find everyone dependent – serfs and lords, vassals and suzerains, laymen and clerics. Personal dependence characterizes the social relations of material production as much as it does the other spheres of life based on that production. But precisely because relations of personal dependence form the given social foundation, there is no need for labour and its products to assume a fantastic form different than their reality.[56]

The economic structure of capitalist society has grown out of the economic structure of feudal society. The dissolution of the latter set free the elements of the former. The immediate producer, the worker, could dispose of his own person only after he had ceased to be bound to the soil and ceased to be the slave, serf of another person. To become a free seller

---

the 'person') is possible only through the establishment of communal social wealth, in a relation that has overcome alienation formed by capital's social relation. Marx writes,
> The less social power the medium of exchange possesses … the greater must be the power of the community which binds the individuals together, the patriarchal relation, the community of antiquity, feudalism and the guild system. Each individual possesses social power in the form of a thing. Rob the thing of this social power and you must give it to persons to exercise over persons. Relations of personal dependence (entirely spontaneous at the outset) are the first social forms, in which human productive capacity develops only to a slight extent and at isolated points. Personal independence founded on objective [*sachlicher*] dependence is the second great form, in which a system of general social metabolism, of universal relations of all-round needs and universal capacities is formed for the first time. Free individuality, based on the universal development of individuals and their subordination of their communal, social productivity as their social wealth, is the third stage (Marx 1973, pp. 157–8).

55   The argument I make here is contrary to the claim made by Kyle Baasch in his article 'The theatre of economic categories: Rediscovering *Capital* in the late 1960s', where Baasch claims that Marx deploys a consistent and systematic distinction between the individual bearer and the impersonal relations of things between persons, or the character mask. I argue instead that there are instances where Marx's person is the person of the impersonal fetish or the mask, and there are times when the person is the bearer themselves (2020).
56   Marx 1990, p. 170.

of labour power, who carries his commodity wherever he can find a market for it, he must further have escaped from the regime of the guilds, their rules for apprentices and journeymen, and their restrictive labour relations.[57]

> The stoical peace of mind with which the political economist regards the most shameless violation of the 'sacred rights of property' and the grossest acts of violence against persons, as soon as they are necessary in order to lay the foundations of the capitalistic mode of production, is shown by Sir F. M. Eden, who is, moreover, Tory and 'philanthropic' in his political colouring.[58]

The use of 'person' as a subject position that might come after capitalist alienation, in which a systematic account would deploy the term 'individual', can be found when Marx defines labour power as a 'personal' form of existence of productive capital, stating in Volume II, 'labour-power [is] the personal form of existence of productive capital', In Chapter 1 of *Capital* Volume I Marx states, 'but every serf knows that what he expends in the service of his lord, is a definite quantity of his own personal labour power'. Taken together, these two statements confirm labour power as internal to an individual person as a transhistorical entity. Labour power in capital's social relations is understood to belong to the 'person' that owns and sells it. However, this 'person' – characterised by the human individual's capacity to labour – is not the same as the commodity owner, but the 'person' is objectified as the thing the commodity owner owns. Here, the person becomes a category for the stuff that makes up the human individual and the empirical reality of one moving and using their body. Therefore, this sense of the person (one's 'own personal labour power') encompasses the living individual and the capacities internal to the life of the individual. As a transhistorical life force, with potential to exist other than it does in capitalism, this use of the person ontologically depicts and requires a position potentially free from capital's domination. This person is thus one of the unrealised individualities that political economy presupposes and also refuses.[59] The person

---

57  Marx 1990, p. 875.
58  Marx 1990, p. 889.
59  Marx refers to the free association of individuals as the world of unalienated communal wealth. This is in line with a more systematic account of the distinction between the individual bearer and the person: Marx does not use the category of the person in this context. However, his account of labour power implies that the category of the person can also

as both presupposed and refused, reflects why Marx does not always deploy a systematic distinction between the 'individual bearer' and 'person'.

Marx often uses the related category of 'man' to denote the ontology of this individual life or person. 'Man' is used by Marx to describe humanity in the terms of an empirical, positivist anthropology that signifies individual humans. To understand the anthropology of this posited presupposition, one requires recourse to discourses that exceed the terms of the critique of political economy, which is itself inadequate to grasp every possible formation of the individual – such as pre-modern or post-colonial individuals. For this reason, although Marx often distinguishes the individual from the person, absent a distinct epistemology that transcends political economy and accounts for anthropological plurality, these concepts used to denote the human individual or person inevitably break down within Marx's critique when placed under of scrutiny.[60]

### 3.2.2   *Person as Juridical Mask*

While Marx does not use the category of the person systematically, the interpretation elicited in this section argues that doing so renders his critique better equipped to address prevailing forms of capitalist domination. A systematic determination of the category of the person is essential if one is to delineate what constitutes 'personal' and 'impersonal' forms of domination.[61] These categories ought to be applied. Doing so enables capital's social relations to be conceived from the standpoint of 'persons' – founded upon abstract right, or the right to own property – as 'juridical masks'.[62] The juridical mask or 'person' is here considered as a performative role that itself grounds the distinction between the mode of expression of an individual (an individual as social medi-

---

stand in to describe the state of the free individual in the context of the production of unalienated communal wealth.

60  The limits of Marx's anthropological presuppositions in his critique is further addressed in Chapter 5, 'Marx's Social Theory of Reproduction', in relation to my argument that Marx has two concepts of life: the life of capital and the transhistorical life of humans and nature.

61  Although facets of 'impersonal' domination, are much taken up in secondary literature – especially within value form analysis – this is not given much attention by Marx. Further, the category of the person as a distinct category from the human individual is given no systematic application apart from its address in *Capital, Volume* I, Chapter 2, 'The Process of Exchange'. Pashukanis and Balibar, among others have subsequently interpreted the person as developed by Marx in that instance to be drawn from Hegel's philosophy of Right. This will be discussed later in this section.

62  Balibar uses this phrase in his essay 'The Social Contract Among Commodities: Marx and the Subject of Exchange' in *Citizen Subject* (2017).

ation) and their mode of existence (an individual's concrete constitution as a bearer of social relations) within social relations of commodity exchange.[63]

Attention to the category of the 'person' supplies an adequate framework to address the various modes of subjection and 'interpellation' of the human individual currently at work within capital's social relations in the context of fictitious capital's increased circulation. This is particularly important to grasp the logical dynamic of 'rentierism'.[64] Moreno Zacarés formulates this problem when he states:

> ... as rentierism takes over, it appears that capitalism's distinct forms of surplus extraction, organized around the impersonal pressures of the world market, are giving way to juridico-political forms of exploitation – fees, leases, politically-sustained capital gains.[65]

Interpersonal forms of exploitation, mediated by capitalism's surplus extraction (exemplifying its impersonal fetish character) are on the rise. These are non-capitalist forms of domination that do not interpellate the human individual through reification but through the 'juridical mask' of the person, which, although set in motion by the exchange abstraction, is not a uniquely capitalist social form.

In the present state of capital's development, the application of a systematic use of the category of the 'person' and the corresponding distinction between impersonal and personal forms of domination are necessary to locate the nature of capital's social form with precision. As such, I seek do so in a way that is philosophically adequate to the respective categories and their role within the logic of capital. Amending Marx's critique with a systematic use of the 'person' as a juridical mask will bestow the tools to understand the forms of domination at work within the circulation of capital and the societal reproduction thereof.

### 3.2.3 Interpersonal Relations

The logical positioning of interpersonal relations within the formal movement of impersonal relations generated by the fetish character of capital's abstrac-

---

63  This distinction anticipates the two concepts of life used by Marx – the life of capital and the life of humans and nature – the former adequately addressed in the discourse of a critique of political economy, the latter requiring a distinct epistemology that tells us what the stuff of life is.
64  Fictitious capital is resultant of the logic of the hoard.
65  Zacarés 2021, p. 49.

tions is exemplified in the phrase 'the personification of things, and reification of persons'[66] [*Personifizierung der sachen und versachlichung der personen*]. In this phrase, the category of the person, as a legal abstraction, or juridical mask,[67] is not the product of the 'reification of the person', and therefore is not exemplary of capital's fetish character. Rather, it is a contrasting abstraction, as a legal category of the person, that is necessary for the possibility of 'personhood's' reification. In this regard, one can distinguish the category 'person' from a living, concrete, sensuous, and potentially free human individual, who is the bearer of this form.[68] As a bearer, the living, potentially free individual is a medium of reproduction of capital's forms.

I.I. Rubin, when he states, 'in capitalist society ... direct relations between determined persons who are owners of different factors of production, do not exist. The capitalist, the wage labourer, as well as the landowner, are commodity owners who are formally independent from each other,'[69] shows that commodities exist in relation to one another through the mediation of individual bearers of the relation, or representatives of factors determined by the production processes. Rubin argues:

> ... separate individuals are related directly to each other by determined production relations, not as members of society, not as persons who occupy a place in the social process of production, but as owners of determined things, as "social representatives" of different factors of production. The capitalist "is merely capital personified." The landlord "appears as the personification of one of the most essential conditions of production," land. This "personification," in which critics of Marx saw something incomprehensible and even mystical, indicates a very real phenomenon: the dependence of production relations among people on the social form of things (factors of production) which belong to them and which are personified by them.[70]

---

66    Marx 1990, p. 209.
67    This term was proposed by Balibar in 'The Social Contract Among Commodities' on p. 199 (2017).
68    As in the case of the commodity form, the living individual endures a non-identity between their concrete sensuous existence and the form in which they appear in capitalist society. This is why, later in the book, I argue that there are two reproductions in Marx: reproduction of life qua life and the reproduction of capital's abstractions which are not entirely distinct.
69    Rubin 2016, p. 22.
70    Rubin 2016, p. 21.

Here, Rubin explains that the bearer of the capital relation, such as the capitalist, is not 'personified' *by* capital; rather, capital, as an object, is personified in its mediation *with* the bearer. This bearer, or living individual, wears the character mask of the capitalist and also the mask of an abstract juridical person as someone with the right to own property.

The living human individual is brought into the logic of commodity and money fetishism by Marx in Chapter 2 of *Capital* Volume I, 'The Process of Exchange'. There, Marx reminds the reader that 'commodities cannot themselves go to the market and perform exchanges in their own right. We must, therefore, have recourse to their guardians, who are the possessors of commodities'.[71] A guardian is individualised as bearer of the capital relation through multiple levels of abstraction, including that of the person.[72] In 'The Process of Exchange', Marx makes explicit that the 'juridical mask' of the 'person' is an essential component of the exchange relation where a legal relation (a contract, which is the relation between two independent 'wills') appears along with the exchange between owners of private property. Marx tells readers that individuals, or 'guardians of commodities', place commodities in relation to one another as persons with abstract will mediated by commodities. Guardians of commodities,

> ... as persons whose will resides in those objects ... must behave in such a way that each does not appropriate the commodity of the other as owners of private property. This juridical relation, whose form is the contract, whether as part of a developed legal system or not, is a relation between two wills which mirrors the economic relation. The content of this juridical relation (or relation of two wills) is itself determined by the economic relation.[73]

Significantly, these juridical persons have 'wills', a phrase used by Marx that indicates that the person is founded upon Hegelian categories of abstract right and therefore retains an element of 'free personality'. For Hegel,

> personality essentially involves the capacity for rights and constitutes the concept and the basis (itself abstract) of the system of abstract and

---

71   Marx 1990, p. 178.
72   Other levels of abstraction in which an individual bearer engages with the capital relation include nationality, gender, race, language, etc.
73   Marx 1990, p. 179.

therefore formal right. Hence the imperative of right is: "Be a person and respect others as persons."[74]

Abstract right is not only a right to own property but also ensures equal rights between persons. As an owner of a commodity, the individual within capital's social form acquires legal subjectivity: they are a bearer of rights, where 'free will' appears in exchange. For Marx, this appears as a contract, where one might find the merger of two wills. Pashukanis also concurs when he states, 'the category of the legal subject is abstracted from the act of exchange taking place in the market. It is precisely in the act of exchange that man puts into practice the formal freedom of self-determination'.[75] Someone who can own a commodity is someone with abstract right.

Marx further insists that persons are individualised to be independent of one another as private owners of commodities. He states:

> ... things are themselves alienable. In order that this alienation [*Verausserung*] may be reciprocal, it is only necessary for men to agree tacitly to treat each other as the private owners of those alienable things, and, precisely for that reason, as persons who are independent of each other.[76]

As Étienne Balibar has argued, the development of the commodity form combined with the institution of money requires individuals to appear as juridical persons. Balibar rightly observes that the social dominance of commodity fetishism requires men to progressively acquire 'the quality of juridical persons and recognis[e] each other as such'[77] as a condition for the social dominance of the commodity form (an alienable thing that is owned), which is established through both individual and a collective practice that enacts ownership. For Balibar, this is a 'subject' characterised not by economic social relations and representations but as 'the subject of law, individualised and represented in the language of juridical institutions'.[78]

As I have shown, Marx understands a contract, which assures the possibility of exchange, to be the relation between 'two wills'. Yet, these are not absolutely

---

74  See section 36 (Hegel 1957).
75  Pashukanis 1989, p. 117.
76  Marx 1990, p. 179.
77  Balibar 2017, p. 190.
78  The full quote reads: 'men, or, more exactly, men progressively acquiring the quality of juridical persons and recognising each other as such, in order for the commodity form to become socially dominant through a specific *action*, an individual and collective practice, what Hegel called *ein Tun aller und jeder*. Herein resides a new characterization of

free wills but wills acting within the framework of capital's social form where persons are representatives of commodities, as explained by Marx:

> ... persons exist for one another merely as representatives and hence owners, of commodities ... [T]he characters who appear on the economic stage are merely personification of economic relations; it is as the bearers of these economic relations that they come into contact with each other.[79]

Therefore, the juridical person is implicit in the exchange abstraction that retrospectively grants commodities their specific form. Pashukanis claims that Marx's analysis of the commodity form is directly followed by an analysis of the form of the subject, and furthermore claims that 'capitalism is a society of commodity-owners first and foremost'.[80] Commodity owners have respective 'wills' that are unified in the exchange. This exchange is conceptualised as contractual as a unity of two wills of respective owners of property: 'a legal-subject – acquires, in compensation as it were, a rare gift: a will, juridically constituted, which makes him absolutely free and equal to other owners of commodities like himself'.[81] Balibar's analysis deepens Pashukanis's observations here when he writes, 'Marx introduces the juridical categories of the *person* (property owner) and the *contract* (unity of wills) as the correlative of the "reflection" of the economic categories of exchange, without which exchange could not occur'.[82] Exchange requires the retrospective positing of the legal person that is abstractly free and equal; however, it is at the same time subject to forms of abstract domination that create practical inequality. Exchange requires a bearer of the relation to be both subject to and subject of the relation. Pushukanis's analysis corresponds with this position when he explains that property is the basis of the legal form only if it can be freely sold on the market. Correlatively, the wage worker 'enters the market as a free vendor of his labour power, which is why the relation of capitalist exploitation is mediated through the form of the contract'.[83]

---

the "subject," this time no longer as bearer of economic relations and their corresponding representations, but as the subject of law, individualised and represented in the language of juridical institutions' (Balibar 2017, p. 190).
79   Marx 1990, pp. 178–9.
80   Pashukanis 1989, pp. 111–12.
81   Pashukanis 1989, p. 114.
82   Balibar 2017, p. 190.
83   Pashukanis 1989, p. 110.

Freedom of will is a significant component of Marx's category of the person. 'Free will' is necessary for capitalist exchange, where contracts are freely taken up and generally not enforced through extra-economic violence. Marx explains the specificity of the capital relation towards the end of *Capital* Volume I in Chapter 28, drawing a distinction between economic and extra-economic domination:

> ... the organisation of the capitalist process of production, once it is fully developed, breaks down all resistance. The constant generation of a relative surplus population keeps the law of the supply and demand of labour, and therefore wages, within narrow limits which correspond to capital's valorisation requirements. The silent compulsion of economic relations sets the seal of the domination of the capitalist over the worker. Direct extra-economic force is still of course used, but only in exceptional cases. In the ordinary run of things, the workers can be left to the 'natural laws of production.'[84]

Exploitation in capitalism takes place within the framework of a contract between free legal individuals. However, to be a legally free and equal bearer (with abstract right) is not akin to the possession of material freedom and equality. Different class positions relate to abstract freedom in ways that are unequal. For example, workers are contradictorily forced to sell their labour freely because it is the only commodity they own.

For Marx, a legal person contrasts to a 'thing', which is the negation of a person as a 'not person' who lacks potential rights or will.[85] Hence, for capital to occupy the place of freedom as an automatic subject, it must be 'personified'. The category of the person is what abstractly renders subjects capable of owning commodities based on the abstract legal capacity to 'freely' dispose of that which they own on the market. Marx reconstructs debates regarding freedom and domination, and sovereignty and subjection, within a theory of capital and, therefore, within a critique of modernity. As a juridical mask, the person retains an element of abstract freedom attributed to the legal category of the 'free per-

---

[84] Marx 1990, p. 899.
[85] The modern use of the word person is derived from the Latin *Persona*, a term for an actor's mask, which is used to identify the difference between the character played and the actor themselves. This term evolved into a double semantic extension through its transformation within modern languages, belonging to grammatical registers on the one hand and the domain of law on the other, where a person is opposed to a thing. See the 'person' entry in *Dictionary of Untranslatables* (Cassin 2014, p. 772).

sonality'. This is so even if the person exists within abstract social relations that dominate through reification, while things – if personified – move with quasi-freedom. The subject of capitalism is therefore not straightforwardly subject to capital but is also a subject of capital's social form. This juridical person is an individual with a 'will', which is, in exchange, enacting a contract or unifying two wills. This is a subject with relative freedom or freedom to act within bourgeois society, expressing Hegel's abstract right.

Commodity ownership as the crux of social relations presents the individual with the conditions for transformation from a living individual bearer to a legal person. Pashukanis explains this process when he outlines the transformation wrought by commodity ownership. He writes that commodity ownership changes

> ... a zoological individual into an abstract, impersonal legal subject, into a legal person. These real conditions are the consolidation of social ties and the growing force of social organisation, that is, of organisation into classes, which culminates in the 'well ordered' bourgeois state.[86]

Pashukanis goes on to explain that within the context of commodified social relations of the bourgeois state

> ... the capacity to be a legal subject is definitively separated from the living concrete personality, ceasing to be a function of its effective conscious will and becoming a purely social function. The capacity to act is itself abstracted from the capacity to possess rights.[87]

Hence the 'right' of commodity owners is abstract, unaffected by the social relations that, through the reification of persons, impose abstract domination over the individual's capacity to act. Therefore, the freedom tied to the legal person is freedom to negotiate contracts within the configuration of bourgeois society, which includes the limits of one's own class. The state confirms this, upholding individual rights to private property. In in doing so, the state sustains the unequal distribution of that property throughout different class positions.

Marx's development of the category of the juridical person in Chapter 2 of *Capital* Volume I extends to address legal persons as differentiated yet abstractly rendered equal in the chapter 'The Trinity Formula' in Volume III.

---

86  Pashukanis 1989, p. 115.
87  Ibid.

Here we find distinct juridical positions of persons created by the capital relation reproduced as unequal persons under the guise of juridical equality. Capital creates three groups of persons – capitalists, workers and landowners – each owning three distinct commodities: capital, labour and land. Marx explains,

> ... capital-profit (or better still capital-interest), land-ground rent, labour-wages, this economic trinity as the connection between the components of value and wealth in general and its sources, completes the mystification of the capitalist mode of production, the reification of social relations, and the immediate coalescence of the material relations of production with their historical and social specificity; the bewitched, distorted and upside-down world haunted by Monsieur le Capital and Madame la Terre, who are at the same time social characters and mere things ... [T]his personification of things and reification of the relations of production.[88]

The combination of the 'economic trinity' completes the mystification by producing distinct classes of persons with distinct legal interests, all of which are required for capital's abstractions to circulate (or for capital's valorisation to occur). The three distinct categories of the person are produced by capital's self-movement in order to complete the mystification, containing the contradiction of being abstractly equal and materially unequal. The trinity formula reflects how one, as an owner of commodity, is subject to the exchange abstraction based on what commodity they own. The legal person is free and equal to sell their commodity within the confines of their class position. Building from Balibar's analysis, we could say that the legal person is a 'juridical mask' that 'subjects must wear who fulfil functions prescribed by the social contract of commodities'. The person is not a state of 'disalienation' but is the '*the subject effect of the structure*'.[89] The individual subjected to the capital relation becomes subject through what Balibar refers to as multiple 'figures of subjectivity', both impersonal and personal forms.

### 3.2.4   *Extra-Economic Forms of Domination*

To summarise the argument given in this section, the reification of persons as character masks of the personification of things requires the additional mediation of the juridical mask of the person. The social relations determined by

---

88   Marx 1991, p. 969.
89   Balibar 2017, p. 199.

capital's abstractions individualise and create the juridical category of the person as one aspect of subjection to capital's social relations. The distinction between the personal and impersonal relation within capital is therefore the result of a logic of subjection developed by the logic of capital's fetish character. Interpersonal relations are extra-economic forms of social domination, within capital social relations, which retain the abstraction of the legal form of a person; they thereby give a particular historical meaning to the constitution of the human individual. Personalisation and depersonalisation – or the personal and impersonal – are, therefore, two sides of the same process. Hence, exchange entails a double subjective effect where the bearer is both subject to impersonal relations and rendered a juridical subject – a 'person' – endowed with a level of individual freedom. The former determines the limits of the element of freedom internal to the latter. A juridical person must freely choose to exchange but can only do so as a wearer of a juridical mask within commodified social relations.

Within commodified social relations, these persons are not confined to the process of exchange only. As outlined by Karl Korsch in his response to Pashukanis,

> ... in the bourgeois 'constitutional state (*Rechtsstaat*)' of today, law has spread, in part actually, in part potentially, from its original sphere of regulating the exchange of commodities of equal value to affect absolutely all social relations existing within modern capitalist society and the state governing.[90]

This claim can be derived from Marx's own logic, when the exchange abstraction is socially dominant, as is the case with the commodity form:

> ... things are in themselves external to man, and therefore alienable. In order that the alienation [*Verausserung*] may be reciprocal, it is only necessary for men to agree tacitly to treat each other as the private owners of those alienable things, and, precisely for that reason, as persons who are independent of each other. But this relationship of reciprocal isolation and foreignness does not exist for the members of a primitive community of natural origin, whether it takes an Inca state. The exchange of commodities begins where communities have their boundaries, at their points of contact with other communities, or with members of the latter.

---

90  See 'Appendix: An Assessment by Karl Korsch' (Pashukanis 1989, p. 189).

> However, as soon as products have become commodities in the external relations of a community, they also, by reaction, become commodities in the internal life of the community.[91]

As in the case of the commodity form, as soon as human individuals become juridical persons in external relations, they become so, too, in the internal life of the community. Therefore, human individuals within capitalist social relations, where commodity fetishism is generalised, become individualised as juridical persons, or as individuals who wear a juridical mask. This mediation of the person is essential for extra-economic forms of domination within capital.[92] This includes the forms of extra-economic domination between persons extended to ensure the reproduction capital's social relations where there is no moment of exchange, such as in the case of debtor-creditor relations, leases, contacts, state benefits, kinship relations, etc. Persons relate to one another interpersonally in the realm of social reproduction and in modes of distribution of value and fictitious value (value that does not go through the production process). When the domination of the exchange abstraction wanes, interpersonal forms of domination become more socially dominant. Amid this network of relations, the 'person' is not a source of disalienation, or given freedoms tied to the bourgeois juridical mask; rather, only the living human individual, the bearer, as a sensuous being can embody a disalienated form of life.

A systematic reading of the 'person' as a category referring to the juridical mask, produced by the exchange abstraction as a necessary subjective effect of commodity fetishism, therefore does not evoke the category as one of 'disalienation' as Marx's unsystematic category of the person can do. Equally, the ontological reality of the living individual cannot be derived fully from the role of the 'person'. However, when the exchange abstraction is socially dominant, the category of the person is applied to all members of society, who then engage in interpersonal struggle for means of reproduction of social life as juridical persons engaged in contracts with landlords, creditors, states, spouses, employers, etc. Hence, the category of the person becomes a more significant place of struggle in the reproduction of capital's social relations when there is crisis in the reproduction of production and increased circulation of fictitious capital.

91  Marx 1990, p. 182.
92  Including juridico-political forms of exploitation.

CHAPTER 4

# Time and Schemas of Reproduction

## Introduction

In *Capital* Volume II, Marx shows that the functional formalism of capital's abstractions requires the reproduction of the concrete means of production. The cycles of production are reproduced relationally and entail a tension between the movement of value through its different abstract forms and the social and natural processes subjected to wear and tear (which require renewal). Correlatively, and in relation to capital's circulation, social and natural processes themselves are reproduced. This chapter supplements Marx's presentation by clarifying that his critique of political economy employed two concepts of 'reproduction' that together make up the process of capital's reproduction. One concept of 'reproduction' was used to explain the reproduction of capital's abstract form, the other the reproduction of human life and nature.[1] This chapter examines Volume II to theorise the interaction between the reproduction of capital's social form and the reproduction of concrete life. This is best approached through a focus on temporalities internal to capital's circuits. By deploying temporal concepts, Marx included the reproduction of concrete life as a variable within his analysis, often referring to 'interruptions' that inform the duration of capital's circuits and thereby impose limitations on the reproduction of capital. Analysis of the formal relation between concrete life processes and capital's abstract forms of domination, as they reproduce through the process of circulation, is pursued in order to challenge the domination of capital's social form over concrete life and to assert the priority of the reproduction of concrete life (both human life and nature). The analysis of circulation reveals concrete life to be an immanent and external variable to the logic of capital. Therefore, struggles for the priority of the reproduction of concrete life, conceived here as 'the time of life', implicitly undermine the domination of value forms.

---

1  There is a third concept of reproduction also used by Marx which entails reproduction in the sense of imitation or mimesis. Marx uses this sense of reproduction when he thinks of historical individuals in terms of 'character masks'. The character mask is the imitative representation of a historical individual that functions to grant the bearer a necessarily objective social role within capital's social relations carried out in practice. The mimetic reproduction of the social role of the character mask works to mediate the other two senses of reproduction.

Building on previous chapters, and furnishing further technical detail, this chapter examines the time and schemas of the reproduction of capital through an explanation of Marx's logical presentation of circuits of circulation in *Capital* Volume II. This will demonstrate that Marx's theory of the capitalist mode of production is conceived logically as a process of reproduction.[2] Capitalism is, for Marx, both a system that posits its own preconditions and a productive system consisting of cycles or periodicity. According to Marx's presentation of reproduction in Volume II, constant capital (capital invested in production) and variable capital (capital invested in hiring labour) need to be re-established and renewed within the circulation process, albeit at different rates and following different rhythms and temporalities.

Circulation renews the conditions of production that will repeat if capital's abstract forms continue to move towards self-valorisation. Circulation, as an interruption to the time of production, distributes commodities ripe for consumption, engages the extraction of resources and facilitates investment and infrastructure. The cyclical nature of capital's reproduction, as it moves between production and circulation, entails contradictions when encountering the concrete reproduction of life and nature, rendering reproduction a process that is both dynamic and cyclical. Therefore, Marx uses the economic sense of 'reproduction' in capitalist social relations to understand the renewal of cycles of production mediated by abstractions. Here, contradictory processes and temporalities unfold when the reproduction of capital's abstract form mediates – and is mediated *by* – the concrete reproduction of human life and nature.

These circuits of capitalist reproduction reveal the concrete social appearance of monetary reproduction: the social content of the reproduction of the system is mobilised by money, begetting further money in a cyclical process. The wage relation[3] underpins this dynamic, acting as the premise for accumulation, linking the reproduction of the lives of persons to the reproduction of capital's abstractions. Under capitalist conditions, money (received as wages or accumulated as surplus value) grants access to the material means of maintain-

---

2 Due to the historical specificity of the way capital's abstractions circulate and accumulate value, an analysis of reproduction is needed. This is not because reproduction is ontologically necessary to think of social or natural processes more generally. It is conceivable that we could think of life as generative and not 'regenerative' for example. The notion of reproduction was never thought in economics until Quesnay developed his Economic Table of 1758, while understandings of the biological reproduction of the 'genus' or species can be attributed to the emergence of natural history.
3 Wage and surplus value.

ing and reproducing the lives of historical individuals. In this way, the monetary relation establishes the cyclical – and therefore temporal – form that social reproduction takes.

Within the Marxist canon, reproduction has generally been understood to play a role comparable to primitive accumulation, whereby capital parasitically feeds on non-capitalist externalities, such as unpaid reproductive labour, non-capitalist societies and natural resources.[4] In contrast, I insist that reproduction needs to be conceived dialectically: the reproduction of capital's abstractions creates the non-capitalist conditions upon which they depend and are structurally required to posit. Reproduction is where capitalist forms meet non-capitalist forms. The resulting tension determines subjects and objects, or the experience of freedom and domination. Capital is not merely a parasitic system and does not reproduce itself only by subsuming non-capitalist forms; rather, the contradiction of capital's reproduction is that capital, as a mode of production, internalises non-capitalist social relations and comes to use them to reproduce itself.

This chapter argues that only Marxist form analysis facilitates adequate comprehension of the nature of the social relationships that reproduce capital. Attention to capital's systematic reproduction is increasingly important when greater strain has befallen reproduction in times of crisis in production, accounted for by the empirical increase of the circulation of credit. While credit is not the focus of Volume II, this chapter shows how credit is assumed in Marx's circuits through the variable of 'hoarding', a negative form of circulation explored previously. In doing so, the chapter illuminates the temporality of interest-bearing capital (fictitious capital) within the process of valorisation. I argue that valorisation needs to be understood as a process that occurs within the reproduction schemas, where, as Marx claims in *Capital* Volume III, 'credit mediates and increases the velocity of circulation'.[5] Crucially, reproduction unfolds at the level of circulation and thus in time, which means that reproduction requires the mediation of credit to uphold its structure.

This chapter argues for the need to foreground circulation in a theory of capital's reproduction, directing attention to the commodity-form function of money, credit and capital as subject, as developed in Volume II (in the schemas of reproduction) and Volume III respectively. As un-valorised value,

---

4 Such as in Rosa Luxemburg's model of reproduction in *The Accumulation of Capital* – where capitalist expansion is understood to rely on an external sector – or some of social reproduction theory that focuses on unpaid labour, such as in the works of Silvia Federici (See Luxemburg 2003; Federici 2012).
5 Marx 1991, p. 654.

credit money is not a distinctly capitalist form of abstraction. It thus becomes a significant example of how non-capitalist forms of subjection are necessary for reproduction. Hence, I argue, issues surrounding reproduction cannot be understood by drawing only on *Capital* Volume I and the passages therein that are so often cited in social reproduction theory in regards to the reproduction of labour.[6] I argue that a full comprehension of Marx's theory of accumulation is needed if we are to adequately explicate a Marxian social theory of reproduction. Marx's interconnected processes of reproduction – abstract and concrete – contain different temporalities, methods, standpoints, relationships and logics that mediate one another in and out of circulation and production, requiring scrutiny of all three Volumes of *Capital*.

The above arguments are first developed in this chapter by establishing the role of reproduction within the circulation of capital (facilitated by the money form). I do so in section, 4.1, 'The Circulation of Capital'. Here, I argue that the circulation of capital engenders what Marx called 'capital's life process', a process that relies on a permanent tension with the reproduction of human life and nature. Subsequently, in the section 4.2, 'Interruptions and Differential Temporal Forms Within Capital's Reproduction', I demonstrate how the practical reproduction of concrete life (human and nature) is negatively related to the formal circulation of capital in Marx's account of temporal interruption within Volume II.[7] In subsequent sections, this chapter proceeds to explain Marx's presentation in Volume II, arguing for the need to develop a complete concept of money. Such a complete theory of money, I contend, ought to integrate credit money into its understanding of capital. Marx's account structurally does so – albeit negatively – through the concept of hoarding. Finally, taken together, three negative variables ('interruptions', 'hoarding' and 'human life/ nature') are argued to exist within the circulation of capital as non-capitalist variables that are produced by capital's circuits as necessary mediators of capital's reproduction. Their interaction dictates that forms of subjection that are both capitalist and non-capitalist persist within the movement of capital's reproduction.

---

6   For the most part, social reproduction theorists have developed their positions through interpretations of *Capital*'s analyses of the wage, where reproductive labour begets labour power and, therefore, is implicit in class struggle in a way that is external to the capitalist economy.

7   This accords with Marx's self-understanding of the exposition explained in Volume III, whereby Volume II addresses circulation 'only in relation of determination of forms it produces' (Marx 1991, p. 967).

## 4.1  The Circulation of Capital

The confrontation between life and form lies at the centre of a dialectical understanding of reproduction. The value form – which encompasses the totality of capitalist social relations – cannot reproduce itself from itself. It requires a spatial and temporal non-capitalist externality as an ontological precondition, even if the precondition is immanently derived from capitalism itself as a system that posits its own presuppositions. This is a condition of the medium of reproduction 'money', which facilitates the abstract reproduction of capital's social forms.

As was demonstrated in Chapter 2, section 2.2, 'Money as Money', money is both capitalist and non-capitalist.[8] This double character is essential to money's ability to operate as a formal medium of reproduction, distinct and autonomous from all other forms. Capitalist production is founded on the confrontation of money with commodities as an 'autonomous form of value'. Put another way, 'exchange-value must obtain an autonomous form in money'.[9] This is possible when one commodity materially stands in to measure the value of all other commodities: 'this thereby becoming the universal commodity, the commodity par excellence, in contrast to all other commodities'.[10] This character of money is developed in Volume II, where Marx's functional formalism demonstrates that the relations between capitalist variables and non-capitalist variables become necessary for the reproduction of capital as a whole. In an account that develops through analysis of the timing of capital's circuits, Marx describes how the combined temporality of capital's self-movement and of its interruptions produce differential durations of turnover time. As such, Volume II indirectly explains how non-capitalist variables (formally subsumed through circulation) are necessary for the reproduction of capital over time (and therefore in practice). Nevertheless, non-capitalist variables remain necessarily external to capital, described by Marx as 'interruptions' within the movement of capital.

In explaining the fundamental importance of circulation as that which facilitates the movement of the process of the accumulation of capital as a

---

[8]  'Non-capitalist' variables and forms are developed from the standpoint of capitalism and therefore do not come from 'pre-capitalist' variables; rather, they are non-capitalist variables produced by capitalist forms. In this regard, they are the fetishistic appearances of interpersonal social relationships produced by the impersonal social form of capital. This 'personal'/'impersonal' distinction is philosophically established in this book in Chapter 3, 'The Fetish Character'.
[9]  Marx 1991, p. 648.
[10]  Ibid.

whole, Volume II connects productive and unproductive activity. Volume I and Volume III are separated by the distinction between productive labour and unproductive labour, and without Volume II we cannot interpret the relationship between the two. The separation of the opening and closing volumes accords with the relationship between human life and the creation (and non-creation) of surplus value: different forms of subjection to different value forms are emphasised on each side – productive capital on the one hand and circulating capital on the other. While retaining a commitment to the interpretation of abstract labour as substance of value – where valorisation relies on anticipated realisation in the production process – the interpretation developed in this chapter claims that value-based subjection equally takes place outside of the production process, requiring attention to be paid to circulation too.

Value-based subjection takes place outside the production process in part because the reproduction of capital as a social form requires the reproduction of human life. Human life is necessary for both the abstraction of abstract labour and the subsumptive character of the movement of the value form within social relations. Nevertheless, human life simultaneously imposes limitations on capital's development due to its need to reproduce itself.[11] Within this tension, different kinds of human subjection are at play. Such modes of subjection are internal to the development of capital as it circulates in a self-reproducing manner. This self-reproductive movement is described by Marx in Volume II as the 'life process of capital'.[12] While living labour, or the kind of human life that produces surplus value, is one of capital's ontologically constitutive variables, the analysis of the circulation and reproduction of capital clarifies that this is not the only relationship between capitalist form and human life upholding capital's movement. There are different kinds of subjection operating in relation to the different changing value forms circulating between production, consumption and finance capital.[13] Therefore attention to the reproduction of human life is needed not only to understand the repro-

---

11  This limit is not undermined on the basis of surplus population, for in order for surplus population to decrease the limitations given by an individual labourer, it also has to reproduce itself. Therefore, the reproduction of human life remains a limit.
12  Marx 1992, pp. 235–6; ibid., p. 273.
13  This argument hinges on a commitment to the address of Volume III as coming before Volume II: Expanded reproduction needs banking capital as a variable, as it is but one aspect of the capitalist mode of production as a whole that circulation connects with the other parts. Volume II is littered with references to content addressed in Volume III, reflecting the necessity to establish the processes occurring in Volume III before one can fully understand the mechanism of circulation and the schemas of reproduction developed in Volume II. This is especially important in accounting for the role of hoarding

duction of labour as correlate to variable capital but also to analyse the connection between different instances of subjection resulting from different tensions and mutual determinations between life and social form within society. The conflict between human life and capital is made especially clear upon consideration of the circulation of capital, which is the mechanism behind both the reproduction of human life (in metabolic relation to nature) and the reproduction of capital as social form.

*Capital* Volume II explicates the ontological basis of capital as a social form through its delineation of circulation time's distinction from capital's production process. The 'overall time of circulation of a given capital' is the sum of both circulation and production, reflecting the entire movement of the valorisation of the value advanced.[14] Volume II shows how the capitalist mode of production exists in relation to human life, not only on the premise of the exploitation of labour but also through the mechanism of the exploitation of human life in circulation more broadly. The mechanistic exploitation of human life in productive labour facilitates subsumption, constituting a distinct form of subjection where human life is both consumed and re-posited in differentiated ways within the circulation process. As a process that reproduces its own conditions to suit its own means, subsumption to capitalist social relations allows capital to extract its substance (abstract labour), which is then employed to drive its self-reproduction. The reification of human labour as abstract labour, according to Marx, lies behind that which gives capital, as a social form itself, a life-like character. Capital appears to 'live' through the circulation process. The tension, therefore, between life and form in Marx is not a strict dichotomy but a process of mutual determination. As in the case of the tension between life and form in Hegel's *The Science of Logic*, the two sides develop into categories that contain both form and content. The nature of the content is determined by the form of thinking and vice versa.[15] The movement of the value form in Marx is the activity of form [*formtatigkeit*], indicating a purposiveness of form often described by Marx as a 'self-movement' or 'life process', where form and matter are not opposed in any fundamental way.

The 'life' of capital is driven by and reproduced through internal relations between value forms (commodity, money, capital) which are mediated by exchange. Hence capital's life process is not directly driven by the life substance of labour. It is driven by the alienated forms whereby abstract labour

---

　　　in Volume II, which acts as a placeholder for the role of credit and banking capital in the reproduction and circulation of capital.
14　　Marx 1992, p. 233.
15　　Meaney 2002, pp. 9–10.

provides mediation as the substance of value. Abstract labour is a quantitative derivation (measured in time) of the qualitative aspects of the encounter between human life and capitalist abstraction in social form, which together form the whole of social relations. To analyse this, capital's life process needs to be read in terms of time. That is, in terms of the relationship between the time of life and the time of capital. Volume II indirectly reveals that to analyse capital ontologically time needs to be interpreted beyond its measured form. The sole interpretation of measured time only provides an understanding of the time of capital – which subsumes and redirects different temporalities. Instead, Volume II reveals that time ought to be analysed in terms of the interruptions internal to the measured time of capital that appear on account of the condition of the time of life, which is tied to non-value producing social practices that contain their own temporalities. The time of life embedded in circulating practices – from the speed of lorries to the time one takes to buy groceries – determines the rate of turnover time that Marx refers to as 'duration'.[16] These interruptions comprise the duration imposed by limits to valorisation on account of the concrete makeup and movement of nature and life: human life always exists in metabolic relation to nature. While life remains empirically irreducible to capital's abstractions, capital's self-movement is formal and based on the measured capture of the substance of life, which is established through reproduction and reposited as immanent to capital. The subsumption of the duration of the time of life and the multiple temporalities comprising life-making practices thus serve as the ontological condition for capital's reproduction.

While the extraction of abstract labour accounts for life's contribution to the substance of the value forms, human life's relation to abstract labour alone reveals little regarding the holistic circulation of capital. Within this process, human life bears the capital relationship not merely as variable capital but also as a function of capital in relation to the numerous different forms involved in production and reproduction. In relation to the commodity form, there are productive capitalists, merchant capitalists and consumers; in relation to the money form, there are salaried workers, money capitalists and debtors; and so on.[17] The actions and conditions of reproduction of these lives, too, mediate the movement of capital.

As the circulation of fictitious capital or credit money has increased, analysis has increasingly employed the concept of 'financialisation' to understand

---

16   Marx 1992, pp. 200–4.
17   While Volume I focuses on wage labourers and productive capitalists, Volume II – as a presentation of circulation – includes, if only structurally, the totality of social life.

changing forms and conditions.[18] The orientation pursued by this chapter is different: here, Volume II and the temporality of human life are employed to understand the specificity of the different ways in which human life bears the function of fictitious capital, an appearance of money that secures money's independence as a reproductive form. This renders an interpretation of money that is crucial for understanding the relationship between individual life and the structural mechanism reproducing capitalist social form. Money's independence is secured in its role as credit, or fictitious capital, because – as credit – money is neither commodity nor capital, but money functioning *as money*. This is due to fictitious capital's status as un-realised value: it has not gone through the production process and does not contain the abstract labour necessary to constitute both commodities and capital. This shift in perspective – to think capital's reproduction from the perspective of money's independence as credit money – enables us to see that with the increased circulation of fictitious capital, as un-realised value, the reproduction of human life suffers greater stress.[19]

The following requires an enriched analysis of reproduction's role in Volume II:

1. Examining the increased proportion of finance and the money form specific to reproduction (fictitious capital/credit money) in relation to production.
2. Addressing the corresponding implications both for the reproduction of human life – as subject to these abstract forms – and for the reproduction of life for its own sake, which will always be both internal and other to capitalistic abstraction.

Accordingly, it is necessary to address Marx's use of Hegel's *Logic* in Volume II philosophically. This is because life, life function and organic form – which constitute essential features of reproduction of capital's reified forms – are developed by Marx through the use of Hegel's logical terms.

Human life constitutes an ontological precondition of the capitalist mode of production; it enables the capitalist mode of production. Human life both provides the substance of value that gives capital its 'life-like' character and is

---

[18] If form establishes the structure of the capitalist economy, then historical changes in proportion to the different elements do not undermine the formal structure facilitating capital's reproduction. In this regard, Marx's schematic of the mode of production ought to be read with attention to historically specific proportions of the variables in a given period.

[19] This argument is established in Chapter 1, 'Fictitious Capital and the Re-emergence of Personal Forms of Domination'.

subjected to value forms, making individual lives bear (in subjection) different forms of appearance of capital. Each value form imposes its distinct fetish character and consequently determines the nature of the subjection of the human bearer in different ways. Therefore, a close analysis of Marx's 'functional formalism', or his theorisation of the value form, can best grasp the philosophical meaning of life in Marx.

## 4.2 Interruptions and Differential Temporal Forms within Capital's Reproduction

The relationship between capitalist social form and human life, in its multiplicity of historical and social determinations, needs to be understood as a presupposition retroactively posited in practice. This positing takes place through the mediation of the empirical constraint imposed by the movement of value forms and the corresponding temporality of natural limits which come to determine the time of social practice. Focus on form, therefore, counterintuitively renders analysis better placed to understand the temporality of individual life – because, as Lukács succinctly observed, '"form" is neither subjectively conjured nor objectively imposed; it holds out the possibility for a mediation and even indissolubility of the subjective and objective realms'.[20] Form reshapes life and creates it anew.

Social practice entails differentiated kinds of subjection occurring within the particular mediations of the time of capital and the time of life, from the realm of production to the reproduction of capital in circulation. Human life meets capitalist social form most directly within the reproduction process. This is owing to the nature of reproduction within the system of capital relations, which takes place within the circulation process. Circulation – which facilitates capitalist reproduction – joins the different elements within the whole of the social life together, creating systematic social relations. In their respective connective, circulatory roles, life and (value) form work together as contradictory elements facilitating the production and reproduction of capitalist social form.

Reproduction, therefore, requires us to mediate a series of key processes encompassed by the encounter between life and form in their organic development. In response, a threefold articulation of reproduction, time and subjectivity is needed. This entails a focus on temporality to interpret the repro-

20  Lukács 2010, p. 3.

duction of capital, with specific focus on the subjection of human life, which is the object of Volume II, *The Process of Circulation of Capital*. Here, Marx demonstrates that to understand reproduction we need to understand the tension between the time of organic being and the abstract time of capital. This is because reproduction (which occurs through the process of circulation) connects different temporalities – including both measured, technical forms of temporality and practical concrete temporalities (interruptions to capital's rotation) – by joining elements within capitalist relations. In turn, these elements co-determine one another through the development of the reproductive process. Within capital's circulation and reproduction, plural modalities of interruption form multiple aspects of a single process.

The reproduction of capital's abstract forms, therefore, requires a synchronisation of different times. A multiplicity of temporalities – made up of concrete practices, disjunctive historiography and material finitude – provide the condition for reproduction. This is because reproduction is a mechanism that requires a supplement to the internal logic of its system and therefore a supplement to the temporal logic of capital. For example, labour time is a concrete temporality that is necessarily other to the abstract temporality of socially necessary labour as exchange value. As Massimiliano Tomba explains:

> The time of abstract labour objectivised in exchange-value does not exist without the time of the concrete and particular labour. The time of labour that determines exchange-value also produces use-value, but these two times are not equal. The clock measures the labour time concretely performed in production, while the time of abstract labour objectivised in the same commodity as socially-necessary labour – thus, as exchange value – has a social measure, given by money. The first temporality is measured by the capitalist or by his overseers with the stopwatch in his right hand and the *Principles of Scientific Management* in the left; the second temporality is, instead, regulated on global markets.[21]

Therefore, a concrete multiplicity of conflicting temporalities – such as labour time and 'free time', or the time of social reproduction and consumption – exist as distinct temporalities to the temporality of the abstract circulation of value. Circulation, nonetheless, requires distinct temporalities for reproduction to occur.

---

21    Tomba 2009, p. 103.

The multifaceted temporality of human practice cannot be fully manipulated or captured by mechanisation in measure.[22] Human life thus remains other to capital. As such, it is critically instructive to consider the co-determining multiple temporalities manifest in Marx's use of the term *duration* when looking to understand the role of social practice within the reproduction schemas that together constitute the duration of turnover in capital's reproductive cycles. While Marx deploys the term 'duration' in Volume II to indicate the time it takes for the turnover of cycles, explicitly connected to the periodicity of 'capital's time of circulation', he fails to account for the fact that cycles have duration because of the way in which the abstractions are mediated by a multiplicity of non-synchronous temporalities of concrete matter, nature and life-making practices. Marx deploys the term 'duration' to stand in for the concrete empirical temporality that exists disjunctively with capital's abstractions without full attention to its content. This makes the otherness of that content opaque and difficult to use for resistance against capital's forms.

As I have elaborated, the reproduction of capital depends on a particular mediation of time and social practice. Ultimately, such mediation is an abstract manipulation of *duration*, or the time it takes for concrete practices to accord with and produce the dynamics of circulation, or capital's cycles, as presented in Volume II. Within the sphere of circulation, differential temporalities – otherwise disconnected – are related to one another: Circulation connects temporalities within the production process on the one hand, and social practice on the other, through the self-moving substance of capital. Although circulation facilitates the valorisation process, which imposes measured linear time on social life, the circulation process cannot be reduced to the valorisation process. This is because circulation includes unrealised value and social practices that, while necessary to uphold the circulation of capital, also uphold human life for the sake of life. In doing so, such practices coexist with distinct temporalities. Nevertheless, as will be subsequently developed, life-making practices with distinct temporalties, which function as the medium of reproduction of capital, are also immanently produced by the reproduction of capital's abstract social form, reflecting the particularity of subjection to unproductive social activity within capitalism. What follows will address the formal workings of reproduction, attending to the mediations of concrete temporalities in circulation that reproduce social relations as a whole.

---

22   For example, ability or disability determines one's pace. Furthermore, one may have a number of other people dependent on them for care, changing the amount of time that one can direct to other things, while, phenomenologically, measured time does not necessarily correlate to one's experience of time.

## 4.3 Marx's Presentation of The Metamorphoses of Capital and Their Circuit

Capital accumulation, or the creation of more capital, occurs through the process of capital's realisation. Valorisation, realised in exchange and facilitated by circulation, is a process that unfolds through the interplay of production and reproduction:[23] the renewal of the conditions of production. At the same time, capital's realisation, at the point of exchange, grounds capitalist society in the inversion of subjectivity in objectivity (or its fetish). Realisation generates more capital by extracting surplus value from abstract labour, the substance of value. Abstract labour is the social form of labour, or that which posits the different value forms to be equal as values. At the moment of exchange, labour time is objectified as socialised abstract labour; at that point, all presuppositions for value as a real abstraction are posited. And since labour is the substance of value, all real abstractions are abstractions of productive labour. Valorisation, therefore, is the process that accumulates capital by facilitating its reproduction in circulation. In doing so, valorisation posits its own presuppositions by endowing labour with an exchange value, measured by its time, which is paid for by the capitalist – only partly – as variable capital.

The money form is the medium of circulation and the immediate shape of value as its measure. Therefore, the money form is also the medium for valorisation and the immediate appearance of value and valorised value (accumulated capital). As the immediate appearance of value, or its measure, money is the form that bears the expression of accumulated capital in its appearance as profit, rendered possible only by its universality as equivalent to all expressions of value. According to Marx, only in the shape of money does value possess 'an independent form by means of which its identity with itself may be asserted'.[24] Christopher J. Arthur notes that this is because only in the form of money do we find homogenous entities both at the beginning and the end of capital's circuit.[25] It is only in the money form that we find both the independent existence of value and the accumulation of more value as valorised capital. What is specific to the nature of capital accumulation is the way in which the inputs and outputs of its reproduction function to produce more money from money on account of formal relations of abstraction that contain quantities of value (this makes money its reproductive form).

---

23   Marx maintained that 'every social process of production is, at the same time, a process of reproduction' (1990, p. 711).
24   Marx 1990, p. 255.
25   Arthur 1998, p. 104.

In order to analyse the conditions of reproduction of capital, Marx developed reproduction tables in *Capital* Volume II. These are structured by the circulation of money.[26] Inputs and outputs of reproduction are represented not as physical qualities of use value but as abstract quantities of value. Because the total social product is comprised of money that becomes more money, the abstract quantities are capitalist in nature. Within capitalist social relations, money capital is invested into the means of production as 'constant capital'. Constant capital is then recovered with the sale of the commodities. Fred Moseley, in his reading of the value-formal role of money in Volume II, describes this when he claims, 'all these variables – price, capital, revenue and so on – are defined in units of money and are not derived in any way from given technical conditions of production'.[27] While the explanation of economics in physical terms of inputs and outputs is common to all economic systems, Marx insists that capitalism's historical specificity resides in the utilisation of money to make more money. The reproduction of the social relations of capital in practice entails the reproduction of money as capital.[28] Through its presentation of the circulation of capital, and therefore of the reproduction of capital, Volume II focuses on the process of capital's valorisation as profit, or, in other words, on capital's 'life process'. To show this, Volume II is divided into three parts: 'The Metamorphoses of Capital and their Circuit', 'The Turnover of Capital' and 'The Reproduction and Circulation of Total Social Capital'. It is therefore instructive to consider the role each part plays.

In the first part, Marx examines the transformations of capital from the point of view of its three circuits – the circuit of money capital (M ... M), production capital (P ... P) and commodity capital (C ... C) – followed by an account of the circuits' unity. In the second part, Marx examines the circuit in terms of its turnover, showing how the different components of capital have different turnover rates internal to their circuit – which, in turn, affect the rate of surplus value. Finally, in 'The Reproduction and Circulation of Total Social Capital', Marx shows how these circuits are moments of individuality internal to the movement of the whole of capital as a mode of production. Marx shows that the circuits, when isolated, are reductive figures reflecting only certain moments within the circulation and reproduction of total social capital. Once understood from the point of view of the whole, the circuits are retrospectively rendered to be aspects of a larger mechanism.

---

26  For an analysis of Marx's schemas of reproduction, with which this book agrees, see De'Ath 2018, p. 397.
27  Moseley 2015, p. 160.
28  Moseley's work accords with this argument (2015, p. 162).

Because the reproduction of capital occurs as a process of circulation, interpreting the reproduction of total social capital, or the reproduction of the 'life' of capital, requires sustained analysis of the medium of circulation: the money form and its circuit. Marx refers to the process of capital's reproduction as the life of capital because the capitalist mode of production is purposed towards its own self-reproducing growth. Capital's reproduction is not the reproduction of self-sustaining renewal; rather, reproduction is the production of more capital. The circuit of money capital, as opposed to the circuit of productive capital, offers a clear lens through which to consider capital's ability to reproduce itself in the sense that its reproduction indicates capital's growth.[29] Focus on the circuit of money capital not only provides insight into the reproduction of capital and its realisation but also constitutes a central feature of Marx's exposition of capital's life process, tied to what he means by capital as 'subject'.

In the first chapter of Volume II, Marx clarifies that the money circuit is a special form of the circuit of capital,[30] distinct from the other two circuits, based on four underlying features:

1. Money is the only form that both opens and closes its circuit. As Marx states, money, unlike the other two circuits, 'forms the starting point and the point of return of the whole process'.[31] Money is advanced and, when advanced, is 'money as capital', where its exchange value and not use value is the purpose of its movement.[32] For example, because the money form both opens and closes the circuit (M-C, C-M), the money form is retained in valorisation and accumulation. When M-C represents individual consumption, this underpins the reproduction of the human bearers of the capital relation. In contrast to the money form, the commodity is ejected from the circuit for the purpose of individual consumption, while the money form is retained by the capitalist and reopens the circuit. This represents a purchase on one side and realisation on the other (where money

---

29  As Tombazos has correctly pointed out, reproduction, characterised as tied to growth, is not dissimilar to the normal sense of the reproduction of a living organism. A living organism reproduces itself in the form of a genus, not for the mere maintenance but for growth (Tombazos 2015, p. 127).
30  Coinciding with two other circuits: the circuit of productive capital and commodity capital.
31  Marx 1992, p. 13.
32  As Marx states, 'it is precisely because the money form of value is its independent and palpable form of appearance that the circulation from M ... M¹, which starts and finishes with actual money, expresses money-making, the driving motive of capitalist production most palpably' (1992, p. 137).

form represents accumulated value valorised at the point of exchange). Thus, money is an expression of capital's reproduction: it remains at hand to reopen the circuit.

2. Production acts as an interruption to money's circuit. The interruption of production, however, is a means for realisation of the value advanced. This demonstrates that the production process serves the purpose of increasing value as it appears in the money form.

3. Money is an independent, palpable form of value where we find no trace of use value. It is an abstraction without a concrete referent. To characterise the independence of money's appearance (combined with money's construal as the aim of the process), Marx refers to the process of the movement of advanced capital to valorised capital with a reproductive analogy. Marx claims the process of the accumulation of money capital to be 'money breeding money',[33] placing emphasis on how the reproduction or growth of this form is intrinsically related to its independence from use value as an independent value form.[34]

4. Money is the only form where the end of the circuit necessarily entails an increase of value. As with production, the 'form P ... P does not necessarily become P ... P¹ (P+$p$), while in commodity circuit the form C₁ ... C¹, no value difference at all is visible between the two extremes'.[35] M-M¹, money begetting more money, uniquely reflects the movement from capital in its form of appearance as value (as the starting point) to more valorised capital (as the result). The advance of capital is the means of the process, and valorised capital is the goal. Both the advance of capital and valorised capital are only expressed in the form of money as an independent value form.

Due to the combinatory function of these four features, the money form is independent of production and consumption. The money form possesses an independent appearance of value belonging to the individual capitalist. The appearance of money will proceed both within the general circulation of commodities and outside the general circulation of commodities.[36] This is due to the fact that, in the circulation sphere, both phases that money goes through – M-C on the one hand and C-M on the other – entail a 'functionally spe-

---

33  Marx 1992, pp. 137–8.
34  As previously stated, its use value is that it creates more value, and therefore the use value itself is an abstraction.
35  Marx 1992, p. 137.
36  Marx 1992, p. 136.

cific character'. These phases in capital's movement are differentiated due to their respective determinations. The material content of labour power and the means of production determine M-C. C-M, by contrast, is determined by capital value realised with surplus value.[37] Value departs and returns to the money form, 'the return of money to its starting-point makes the movement M ... M¹ a cyclical movement complete in itself'.[38] The movement of money appears independently from productive consumption, which takes place between the two stages. However, valorisation nonetheless occurs within the realm of reproduction and circulation when M¹ appears, representing the valorisation of value and the 'aim and driving motive' of the process. Each circuit constitutes a section that is repeated within the overall process aimed towards valorising value. As Marx states,

> ... in the life of capital, the individual circuit forms only a section that is constantly repeated, i.e. a period. At the close of the period M ... M¹, the capital exists again in the form of money capital and passes once more through the series of changes of form that constitute its process of reproduction and valorisation.[39]

The exchange of commodities links the production process to the reproduction process by circulating commodities that reproduce the conditions of capital accumulation, either by providing the means of production (constant capital) bought by the capitalist or the means of social reproduction (reproduction of human life), which in turn provide the means of subsistence (consumption). Commodities sold for consumption are both bought by the capitalist and workers. Yet, in the case of workers, the reproduction of the means of subsistence also reproduces labour forces or variable capital (a means of production). Consumption contributes both to the reproduction of value through the reproduction of the variable of labour and to the physical reproduction of human life. Consumption therefore plays a pivotal role in the reproduction of all bearers of different value forms, whether they be direct bearers or indirect bearers, such as in the lives of unpaid social-reproductive labourers.

---

[37] Ibid.
[38] Ibid.
[39] Marx 1992, p. 252.

## 4.4  Marx's Presentation of The Turnover of Capital

In the second section of Volume II, 'The Turnover of Capital', Marx focuses on the duration of turnover times. This includes the sum of production time and circulation time, which together constitute capital's life process. In doing so, Marx elucidates the way in which reproduction occurs in practice. Turnover time is presented as the means for the reproduction of capital advanced, where self-valorising value moves between the forms in practice by appearing as use values. Turnover time begins the moment capital is advanced and elapses with the return of this capital in the same form (money). Marking an essential feature of capitalist production, turnover time is the time it takes for the valorisation of capital that has been advanced.[40] In this section, Marx addresses how the different circuits – the circuit of money capital, the circuit of productive capital and the circuit of commodity capital – possess different durations. Within the turnover of capital, the difference between fixed and circulating capital[41] affects these durations according to their respective cycles. Marx emphasises how capital's metamorphosis necessarily moves in and out of the money form over the course of the turnover because of money's role in expressing value.[42] Here, monetary value, expressed through the accumulation of surplus value, is a consequence of the duration of the different components of turnover time. Arthur explains this temporal structure, stating,

> ... time is of the essence of all economies, but of capital above all. For the whole idea of valorisation rests conceptually on just such a comparison of capital value across time. It is between these times that capital accomplishes its circuit of transformations.[43]

The changes that take place in capital value over time are determined by circulation and its medium, money, as they connect different social practices containing different durations, or the times 'between' value's appearance within the production process. These times between production accomplish capital's

---

40   Marx 1992, p. 233.
41   Note that it is within Marx's analysis of turnover time where the distinction between fixed and circulating capital is important, rather than the distinction between constant and circulating capital developed later when he describes the two departments.
42   Money expressing value is value relating to itself.
43   Arthur 1998, p. 97.

circuit of transformations by facilitating the passage of commodities from production to point of sale or consumption (both productive and unproductive consumption). Marx details:

> The annual product includes both the parts of the social product that replace capital, and what characteristics are common to both. The annual product includes both the parts of the social product that replace capital, social reproduction, and the parts that accrue to the consumption fund and are consumed by workers and capitalists: i.e. both productive and unproductive consumption. This consumption thus includes the reproduction (i.e. maintenance) of the capitalist class and the working class, and hence too the reproduction of the capitalist character of the entire production process.[44]

Within this process, human life is subject to different value forms and not just the labour-power commodity. Hence the duration of human life, or the practical content of human life in capitalist social relations, which is consistently reproducing its *own* life, contributes to the reproduction of capital through subjection to social form. Subjection to social form occurs in ways that are not merely based on productive labour: The subjection of human life to different moments of the circulation process affects turnover time and therefore the movement of capital's abstractions by causing interruptions in circulation. Human life might increase or decrease the time of circulation due to the limits and intervening temporalties imposed by natural requirements and individual variance. However, because this analysis is concerned with the reproduction of total social capital, the turnover times will only be of use after establishing an understanding of reproduction and circulation of total social capital: the totality of relations of the movement of the 'concept' of capital or expanded reproduction.

## 4.5 Marx's Presentation of the Reproduction and Circulation of Total Social Capital

In the third section of *Capital* Volume II, 'Reproduction and Circulation of Total Social Capital', Marx closely analyses the reproduction process as it proceeds through circulation to show how prior misunderstandings of the differences

---

44  Marx 1992, p. 468.

between constant and variable capital led to the misconstruing of profit in classical economics. According to Marx, classical economics did not recognise the asymmetry between constant capital and variable capital. This was due to misrecognition of the fact that capital invested in variable capital only covers a portion of its cost. In order to show that variable capital is systematically only partially accounted for, Marx had to develop a detailed account of valorisation with a grounding in reproduction, a process that both requires and distributes consumption. Marx formulates the problem with the following question:

> How is the *capital* consumed in production replaced in its value out of the annual product, and how is the movement of this replacement intertwined with the consumption of surplus-value by the capitalist and of wages by the workers?[45]

This question implies reproduction and implicates both productive and unproductive consumption.

The distribution of consumption engendering the capitalist relation takes place at the level of simple reproduction. Simple reproduction, the appearance of an isolated function of reproduction, is an alternative conceptual depiction of the process of reproduction as a whole (as is the case of circulation). The wage relation internal to simple reproduction posits the consumption of the worker (variable capital), while the surplus value extracted posits the consumption of the capitalist (who will both invest in fixed capital and engage in individual consumption). These two 'Departments' of simple reproduction already critically expose the false appearance of the wage relation as a free contractual act between two agents (one selling labour and the other buying). The form of consumption distributed by simple circulation is the reproduction of the social form, and it confirms the presupposition that the wage labourer is already dependent on the capitalist: in this way, the simple reproduction of variable capital contains in it the abstract concept of total social capital's relations. However, while simple reproduction might retain the concept of the social relation (entailing a conceptual connection), it does not contain practical concrete instances of non-capitalist mediation or its condition. These latter factors – the concrete nature of what the abstraction means in practice – elude abstract theorisation.

Reproduction posits, or produces, capital's social form. In doing so, it secures the conditions of possibility of production's consumption, which requires

---

45   Marx 1992, p. 469.

ongoing renewal through fixed and variable capital. Simple reproduction reflects this and therefore the concept of the social form inheres within it.[46] The concept of the social form is internal to simple reproduction because the false appearance of the free contract between the worker and the capitalist is revealed by the necessity for the wage relation to reproduce one's life in the first place. The wage, as that which is required for the worker's consumption, is not a result of production but the immanent precondition. Simple reproduction posits the social form that endows the worker with a character mask, making them a function of their class as a bearer of the social form.

In simple reproduction, capital is invested both as constant capital and as variable capital, divided between what Marx calls the two departments of social production: department I and department II. Department I is the *Means of Production*: the department that produces commodities that either enter, or are capable of entering, productive consumption. Department II is the *Means of Consumption*: the department producing commodities that are consumed by individuals (individuals include all classes, the capitalist class as well as the working classes). Variable and constant capital are internal to both departments.

The concrete makeup of the commodity, or its content, opposed to the abstraction of its value form, only becomes important in Marx's exposition when considering capitalist relations as total social capital. This is because it is only when thinking of capitalist social form from the point of view of total social capital (the conditions of possibility for the accumulation of value) that it is necessary to account for reproduction (both simple and expanded). Taken individually, the specific commodities produced remain indifferent for Marx's analysis, since what constitutes the commodity – its concrete content – is important only for the reproduction of the system. Marx explains,

> As long as we were dealing with capital's value production and the value of its product individually, the natural form of the commodity product was a matter of complete indifference for the analysis, whether it was machines or corn or mirrors.[47]

Individual capital is interested in the value of its product individually, not the replacement and consumption of materials. The replacement and consumption of materials, however, are two significant processes of capital's reproduc-

---

[46] For an expansion on this reading, see Étienne Balibar's contribution to *Reading Capital*, 'On the Basic Concepts of Historical Materialism' (Althusser et al. 2015).
[47] Marx 1992, p. 470.

tion, and of social reproduction more generally, which do not entail indifference to the concrete.[48] Within the movement of total social capital, values and concrete material of the use values are replaced. Marx continues, stating that the movement of total social capital and the value of its product 'is not only a replacement of values, but a replacement of materials, and is therefore conditioned not just by the mutual relations of the value components of the social product but equally by their use-values, their material shape'.[49]

Reproduction occurs as a process comprising the interrelation of two departments, where the replacement of use values, or concrete material, conditions the replacement of values. The abstract value components of the social product determine individual capital. But without the specificity of the concrete material, they cannot reproduce total social capital and thus cannot reproduce capital as a social relation. Therefore, the process of valorisation occurring between these departments is developed as a theory of reproduction. Reproduction as a process is unique in the critique of political economy, in that capital's abstractions depend on the specificity of the content of the concrete. This theory of reproduction, or valorisation occurring between the two departments, is used by Marx to expose the fetish character of profit, which 'appears' to be gained through elements of supply and demand. Profit, the creation of more capital, is the form of appearance of capital accumulation, or capital's growth. Marx's theory of valorisation (a theory of reproduction between the two departments) undermines the appearance of profit as articulated in classical economics. Marx does this by showing how the reproduction of capital facilitates accumulation through the exploitation of variable capital.

Variable capital, the cost of hiring labour, represents abstract labour, or labour as measured in value. These wages only cover a portion of working time and, as a result, the capitalist keeps the value of abstract labour not paid in wages as surplus value. Abstract labour as the source of new value – represented by the value forms – constitutes the substance of value; it is the content of the abstraction. Marx analyses this problem through consideration of substance and form. By so doing, he is able to explain a dynamic occurring beneath the appearance of wage, profit, price, interest, rent, and so on. When abstract labour, as the substance of value, appears as a wage, profit, or price etc., classical political economists took these categories to represent value immediately, without thought to their underlying substance or appearance as inverted social forms. By thinking this problem as a problem of substance and form,

---

48  Since capital's abstractions are by necessity indifferent to its content, reproduction requires a mediation that is non-capitalist.

49  Marx 1992, p. 470.

Marx explains that abstract labour, as substance, itself cannot but appear other than as a distorted form.[50] It is from this perspective that the valorisation process – which takes place on the basis of the movement of the value form, whereby essence appears in inverted form – takes its method from Hegel's *Logic*. The concept (capital) has to move through different moments (reflected in its forms) to develop into valorised value and capital as subject/idea/concept (capital that generates more capital as a self-moving process).

Because *Capital* is a critique of political economy as a science, comprehension of the scientific method deployed by Marx to undertake his critique is central to understanding what the critique achieves.[51] Marx's method, deploying a critique of political economy through the use of Hegel's systemic logic, which pursues totality, is clearly presented in his expositions of reproduction and the reproductive schemas. Marx's focus on revealing the fetish character of profit presented by classical political economy is combined with an analysis of reproduction not from the point of view of simple reproduction but from that of expanded reproduction: the reproduction of the capitalist system as a whole. Expanded reproduction includes various spheres of circulation, production and banking capital. When taken together, the analysis of these spheres reveals how production and circulation are interlinked in such a way as to account for the production of surplus value. It is precisely this dynamic that constitutes capital's logic. Nonetheless, following Marx's method, the exposition of simple reproduction is a necessary moment of explanation, first appearing as a conceptual abstraction that will later engender a better understanding of expanded circulation.

'Simple reproduction on the same scale' is an abstraction because it entails an absence of accumulation or reproduction on an expanded scale, where concrete inputs or conditions of production change over time.[52] Nonetheless, as an abstract social relation, simple reproduction contains the concept of reproduction. By looking at simple reproduction, which is distinct from the valorisation of individual capital as one moment of the movement of capital, Marx reveals what lies behind the appearance of capital. As a unity of different moments of both circulation and production, capital must appear as simple reproduction. When appearing as simple reproduction, however, capital appears to have obtained its own 'self-movement' as a presupposition for the unity of production and circulation. However, these two sides – production and circulation –

---

50   See Lange 2016, p. 153.
51   For a detailed commentary on Marx's application of the Hegelian scientific method, see Lange 2016, p. 151.
52   Marx 1992, p. 470.

are transformed by capital when developing as an 'end-in-itself', as money begetting money, in contrast to simple reproduction, which, while depicting a social relation of dependency, does not require the accumulation of profit in its concept.[53]

## 4.6   The Three Circuits of Capital

In his presentation of circulation in Volume II, Marx begins by presenting the three circuits of capital. The circuit of money capital, productive capital and commodity capital are represented by the following formulas:

Money capital:      M-C ... P ... C¹-M¹
Productive capital: P ... C¹-M¹-C ... P
Commodity capital:  C¹-M¹-C ... P ... C¹

In all cases, the product is also the premise. Taken together, where Tc stands for the total circulation process, and the dots refer to interruptions, qualified by duration, we have the following:

(I)   M-C ... P ... C¹-M¹
(II)  P ... Tc ... P
(III) Tc ... P(C¹)

When understood as functioning together, all three forms are made up of premises that are results produced by the process. Marx explains,

> If we take all three forms together, then all the premises of the process appear as its result, as premises produced by the process itself. Each moment appears as a point of departure, of transit, and of return.[54]

The total process – where each circuit is either a point of departure, transit or return for another circuit – is the unity of production and circulation. Production is a mediator of circulation and circulation a mediator of production. Each circuit holds in common the fact that value's valorisation is its purpose, or, in

---

53   For a more detailed discussion of the transformation of circulation and production for the purpose of capital's movement as an 'end-in-itself', see Meaney 2002, p. 11.
54   Marx 1992, p. 180.

Marx's words, value's valorisation is 'the driving motive' of each circuit.[55] The form of figure (I) expresses this in that more money is accumulated. In figure (III), the circuit opens with value that is already valorised and closes with new valorised value (even when the circuit repeats on the same scale).[56]

The repetition of one circuit includes variables that necessarily take place in each respective circuit. In this way, 'the entire circuit is the real unity of the three forms'[57] that constitute an equal number of interruptions. The circuit of productive capital is not only a 'periodic renewal' of productive capital, but it is also presented as an interruption in the production process whilst the circulation process continues. Production is not continuous but 'pursued only in spasms to be repeated only after periods of time of accidental duration, according to whether the two stages of the circulation process were accomplished quicker or more slowly'.[58] The duration of turnover times comprise the temporality of the different moments within the circuit that correspond to the duration of empirical movement, such as the transportation of commodities, the growth of natural resources and the performance of roles taken on by human subjects (who must reproduce their own lives within this sphere of circulation).

Of all three circuits, the money circuit is the only circuit that gains more value in its closure. In this regard, C-M-C (including both productive circuits, P ... $C^1$-$M^1$-C ... P, and commodity circuits, $C^1$-$M^1$-C ... P ... $C^1$) is put to work for the purpose of M-C-$M^1$, where the increase of M is the aim of the process. C-M-C, the process of simple circulation, is determined by the circuits of capital, making the process socially and historically general only in relation to the dominating process of M-C-$M^1$, which is occurring at the same time. In this manner, C-M-C is not characteristic of capital as a mode of production, while the process of M-C-$M^1$ plays a dominant role in history.[59]

## 4.7  The Role of the Credit System within Capital's Reproduction

With these circuits in mind, it is imperative to consider the credit system and credit money. Such consideration is required to accurately account for the valorisation process (and therefore for the reproduction of capital). Both the

---

55   Ibid.
56   Ibid.
57   Marx 1992, p. 181.
58   Marx 1992, pp. 181–2.
59   For further discussion on M-C-M as the dominating process, see Tombazos 2015, p. 80.

credit system and credit money play a significant role in capital's life process, and both subject individuals to the circulation process. It is important to bring this component, an aspect of a complete theory of money, into a thoroughgoing analysis of Volume II because the durational interruptions within circulation are implicated in the use of credit money.

Because Volume II appears before Volume III, it is often assumed that credit operations and banking capital do not play a meaningful role in Volume II. However, if we look at the function of hoarding in Volume II, it becomes apparent that such assumptions are mistaken. Hoarding is the foundation of the credit system and is a significant requirement for circulation, consistently appearing throughout its movement. The foundation of Marx's explanation of the credit system rests on his insistence that money hoards are necessary for circulation to function. Further, by transforming their hoards into interest-bearing capital, capitalists are able to gain an increased share of surplus value. The upshot of this is a concentration of money capital in banks, bonds and stock markets. Hoarding, which I have termed *negative circulation* in subsequent chapters, accounts for the variable of banking capital that is later developed in detail in Volume III. The credit system necessitates a more complex presentation of capitalist reproduction; it therefore obliges an analysis of Volume II from the point of view of Volume III, whereby credit can be understood in its more developed form to account for the reproduction of capital more fully. This is because both productive and unproductive social relations are reproduced through the movement of the value form and valorisation, a movement that does not exist in isolation from the credit system. Because Marx utilises hoards to abstract from the credit system in Volume II, the description cannot accurately capture the reality of capital's reproduction. Not only do the hoards not actually exist in the way that Marx describes, but it is in fact not his claim that they do. In reality, money hoards are utilised as financial assets in a number of ways, such as in bank accounts. These funds are subsequently used by some capitalists and lent out to others. The variable of the hoard is a sum of money in circulation, circulating negatively or positively, outside of the production process, playing an important role in the formal dynamic of the reproduction of capital.

While credit money is an essential variable of reproduction that contributes to the interruption of valorisation, it simultaneously continues to be an agent of reproduction both in relation to the reproduction of human life and the reproduction of the life of capital. Credit money does so through interest-bearing capital (fictitious capital). Interest-bearing capital determines the empirical duration of valorisation (capital's life process) and is present already in Volume II, often disguised as hoarded money. As Marx states, capital's life process is 'the time required for the renewal and repetition of the valorization and production process of the same capital value'.[60] Therefore, if capital's life process is an abstract process of renewal and repetition that engages concrete lived renewal and repetition, a robust analysis ought to include all aspects that contribute to the circulation of this process to obtain an acute understanding.

Marx, having written the manuscript for Volume II five years after the manuscript for Volume III,[61] was well aware of the need to include the credit system in his analysis of the circulation of money in the reproductive schemas. As Marx states,

> ... if we consider [reproduction] exclusively on the assumption of simple money circulation, without any regard to the credit system (this will be brought in later), then the mechanism of the movement is as follows. In the first volume (Chapter 3, 3, a) it was shown that although part of the money present in a society always lies fallow in the form of a hoard, while another part functions as means of circulation or as an immediate reserve fund of directly circulating money, the proportion in which the total quantity of money is divided between hoard and means of circulation is consistently altered ... ... With the development of the credit system, which necessarily runs parallel with the development of large-scale industry and capitalist production, this money no longer functions as a hoard but as capital, though not in the hands of its proprietor, but rather of other capitalists at whose disposal it is put.[62]

This statement clarifies that a full picture of the circulation of money requires the inclusion of interest-bearing capital, which is not a further development or higher category of the process, but rather a necessary component of the internal mechanisms of reproduction.

## 4.8    Expanded Reproduction

In order for accumulation to take place, capital requires credit money, which only circulates within expanded reproduction. Simple reproduction must be concretely presupposed by total social capital, while the total social capital relationship is posited by simple reproduction. Therefore, although simple

---

60    Marx 1992, p. 236.
61    Arthur and Reuten 1998, p. 9.
62    Marx 1992, p. 261.

reproduction contains the concept of and posits capital's social form, capitalist social relations and non-capitalist relations – which together make up the social totality – cannot be understood in terms of simple reproduction only. An analysis of capital requires analysis of expanded reproduction, which encompasses the multifaceted model presented across all three volumes, where banking capital and interest-bearing capital are necessary variables.

The distinction between simple and expanded reproduction is established in Volume I. In the first book of Volume I, the 'Process of Production of Capital', Marx clarifies that capital is a cyclical process.[63] Marx explains,

> ... in simple circulation, the value of commodities attained at the most a form independent of their use values, i.e. the form of money. But now, in the circulation M-C-M¹, value suddenly presents itself as a self-moving substance which passes through a process of its own, and for which commodities and money are both mere forms ... [V]alue therefore now becomes value in process, money in process, and, as such, capital.[64]

Volume I nonetheless focuses on a hypothetical structure where the scale of production is consistent, which Marx has abstracted from circulation. While capital's social form is posited, accumulation is not. Simple reproduction appears in this way because Marx was assigning a starting point for his exposition, where the categories or forms presented are already socially mediated. Thus, Marx's starting point must be interpreted as predetermined by conceptual presuppositions. Marx's initial articulation of the value forms are intentionally false (in the sense that it is partial) and already mediated by the social relations that underpin them. The presentation must be false because the categories are presented in their isolation as conceptual presuppositions. These conceptual presuppositions mediate the unmediated categories in simple reproduction and are understood by the end of Volume II as based on the mediation of expanded reproduction: a necessary mediation for every reproduction. This shows that expanded reproduction is necessary for capital accumulation to take place concretely.

Simple reproduction exists in the moments in which circulating capital (social capital) is posited in a 'specific determination' (a Hegelian term for the negation of the movement from the universal to the particular).[65] As Marx explains, 'while capital as the totality of circulation is circulating capital, the

---

63   This point is developed in Arthur 1998, p. 96.
64   Marx 1992, p. 256.
65   Arthur, Christopher, J. in ed. Arthur and Reuten

transformation from one phase to another, it is, in each phase, also posited in a specific determination, confined to a particular form, which negates it as the subject of the movement as a whole'.[66] This demonstrates that while simple reproduction is a part of expanded reproduction, the inputs and outputs – and their proportion – rely on the dynamic of total social capital. In this regard, the schemas of expanded reproduction illuminate the unity of the processes of production and circulation, clarifying their role as foundation for other concrete relations, including finance and banking capital. The schemas, too, bestow the tools for a complete concept of money (on the basis of the explicit explanation of hoarding).

Before Marx's presentation of the production of capitalist commodities, Volume I offers an analysis of 'simple commodity circulation'. To get to the sphere of production, readers have to reach the end of Part Two of Volume 1. Only then do they leave the sphere of circulation (where what takes place does so on the surface 'in full view') and enter the 'hidden abode of production', where they discover how capital produces and is produced.[67] This order of exposition was employed because production of capital is production of value, and value is a form constituted in exchange, which occurs in circulation.[68] In contrast, Volume II is focused on the 'social circulation of capital'.[69] Time spent within the production process is included in social circulation, which is addressed in part in Volume I. Volume II of *Capital* reconceptualises the presentation of circulation as it appears in Volume I. It does this through an examination of circulation, exceeding its conceptualisation as the circulation of commodities to develop its conceptualisation as a bearer of the circuit of capital. Far from merely enabling the movement of commodities, capital as a circuit subsumes money and commodities in the movement towards valorisation.[70] Marx introduces the phrase 'the life of capital' at this point. He does this because capital is shown to be a subject that contains within itself a *drive for valorisation* through its process of social circulation.

The presentation of circulation that takes place in Volume II entails the reconceptualisation of money and commodities as more than mere surface phenomena. Instead, they are presented as 'forms of capital's self-positing movement'.[71] Capital goes through a series of metamorphoses, repetitively

66  Marx 1973, p. 621.
67  Marx 1847, pp. 279–80.
68  See Arthur and Reuten 1998.
69  Arthur and Reuten 1998, pp. 3–4.
70  Ibid.
71  Arthur and Reuten 1998, p. 5.

moving through the phases of being money capital, commodity capital and productive capital. In Volume II, Marx reworks what was described in Volume I with a level of conceptualisation that is more comprehensive and therefore more concrete.[72] It is for this reason that, in the context of Volume II, turnover time requires both a discussion of time spent in production and time spent in circulation on the market: two sides internal to the social circulation of capital.[73]

Simple reproduction in *Capital* articulates an isolated structure where surplus value produced in the two departments of reproduction (production of means of production and the production of means of subsistence) are exhausted by the consumption of the capitalists. As Amy De'Ath has observed, 'simple reproduction describes a situation in which all surplus value produced in both departments is "used up" or personally consumed by the capitalists'.[74] Here, money is not advanced but is spent. For reproduction to be simple, there will need to be adequate proportionality between the two departments: constant capital (capital produced for means of production) and variable capital (capital produced for means of consumption). However, due to the complexity of circulation time and the need to finance the means of production among other factors in practice, adequate proportionality is not necessarily obtained. Yet at the level of individual capital, for the production of both value and output to be the same each year, the relation between each department must be proportional.[75] Expanded reproduction, with its focus on social capital, allows us to see that individual capitals have different turnover times and rates of consumption of fixed capital. These operate as restraints preventing proportionality between the two departments in practice.

### 4.9 A Complete Concept of Money for Understanding Capital's Reproduction

Marx's schemas of simple reproduction are missing a complete concept of money. This prohibits readers from grasping expanded reproduction or the reproduction of capital's social relations as a whole, a process that requires the specificity of the money form in its complete presentation. As a means to reduce both purchasing and selling time, and therefore to reduce value lost in

---

72   Ibid.
73   Ibid.
74   De'Ath 2018, p. 397.
75   Tombazos 2015, p. 209.

circulation, the credit system helps to simplify the mobility of industrial capital.[76] The simplification of the mobility of capital is reflected in Part Three of Volume II, 'The Reproduction and Circulation of Total Social Capital', where Marx presents credit money as a means of financing production. This increases the mobility of production not only by reducing circulation time, but also by facilitating the growth of unproductive sectors. Thus, expanded reproduction needs the credit system and finance capital. While this is marginally addressed in Volume II, credit and finance retain the status of a variable, largely in the place of what Marx refers to as hoards when describing the circuits and reproduction schemas. As a source of funds in the credit system, hoards are necessary for capital's circulation. Therefore, disruptions, or the lack of disruptions in capital's circulation, are determined or elevated by the conditions of the credit system.[77] This reflects how simple circulation presupposes the formation of money hoards. Hoards, although an interruption in the ongoing movement of a circuit, are a requirement for future circulation. There must be a discontinuity in spending on the basis of value accumulated in independence from productive capital (in hoards / the credit system).[78] As Marx explains,

> We showed in Volume I how accumulation proceeds for the individual capitalist. The realization of his commodity capital also brings with it the realization of the surplus product in which his surplus-value is represented. The surplus-value that is transformed into money in this way is then transformed back by the capitalist into additional natural elements of his productive capital. In the next production circuit, the increased capital supplies an increased product. But what occurs in the case of an individual capital must also occur in the overall annual reproduction, just as we have seen that what in the case of the individual capital is the successive precipitation of its worn-out fixed components in money that is hoarded up, also finds its expression in the annual social reproduction.[79]

When tracing the role of the money form, it becomes apparent that the circulation of capital must include an element of surplus value accumulated in interruption from productive capital (as a necessary component of social reproduction). This is not only because interest-bearing capital plays a role in supplying money in advance, but also because its role can only be understood

---

76  Tombazos 2015, pp. 256–7.
77  Martha Campbell in ed. Arthur and Reuten 1998, p. 130.
78  Martha Campbell in ed. Arthur and Reuten 1998, p. 129.
79  Marx 1992, p. 565.

from the point of view of social totality (expanded reproduction). There, the components of capital emphasised in Marx's circuits do not mirror each other symmetrically.

While productive capital is essential for supplying abstract labour, other circuits of capital are not always directly posed in relation to it: the commodity disappears from the circuit in consumption, while money retains its independence from production when accumulated in hoards. It is the special role of credit (derived from hoards) that gives the money form its function as medium of circulation, because – due to the asymmetry of the circuits – there needs to be a universal equivalent that can mediate disjunctive time. In this way, quoting de Brunhoff, whose work helped to establish my argument in section 2.2,

> ... money as a financial instrument preserves its specific characteristics as a non-commodity ... [T]he role of the monetary relationship when it has become "immanent" in capitalist reproduction. In simple circulation, money as general equivalent is distinct from all the commodities exchanged by private producers. In the circulation of capital "money-capital evidently plays a prominent role, seeing that it is the form in which the variable capital is advanced," insofar as "the wage system predominates." The use of the money is thus doubly determined by the social relationships between private economic agents. But money still remains true to its nature, witness the financial role of hoarding.[80]

Here, de Brunhoff correctly asserts that money remains true to its independence in order to facilitate the passage of value between disjunctive circuits. In so doing, money secures its role in financing reproduction; this is evident in hoarding. Only in expanded reproduction and its circulation, which leads to valorisation in the reproduction of social totality, is the development of all three circuits articulated relationally. Expanded reproduction relies on credit as a connecting agent behind the non-synchronicity of the circuits. This is possible because hoarding allows money to be used as credit without dependence on the timing of the realisation of value determined by the production process.

A satisfactory analysis of the social totality – or unity of the system – therefore requires analysis of the conditions of the commodification of labour. The capitalist mode of production reproduces itself within generalised commodity exchange on the basis of the movement of the value form, where the credit system remains a necessary component. Furthermore, social totality is the frame

---

80    de Brunhoff 2015, p. 71.

from which capital's manifestation in the final moment in the development of the concept (capital) into the Hegelian categories of 'Idea' or Concept/Subject[81] can be conceived. Hence, in reconceptualising circulation as circuits of capital's self-movement, the expanded reproduction schemas play an important role in the development of the categories of substance, subject and life.

Capital's self-movement is grounded in human life, which – as a presupposition for abstract labour – forms the social 'substance' of value. Marx's labour theory of value reveals the fetish character of the bourgeois relations of production, which rely on the truth of appearances without grasping the essence of appearances. The independent movement of value between these forms of appearance can only be 'the force of [an] elemental natural process'[82] because value is derived from living labour. It is this rootedness in life that, following the exposition of Hegel's categories in the doctrine of the concept, evokes a concept of subject. Valorisation hinges on the self-movement of capital throughout the value forms. However, it is only when analysis moves to the 'idea', within the development of the concept, that it can adequately produce its movement of self-valorisation.

### 4.10   Non-capitalist Variables within Capital's Reproduction

That which reproduces capital must supplement capital's own logic as an internal logical necessity. Lange has observed that 'as a general logical prerequisite, it is obvious that what *accounts for* an entity must be of a different category and quality *than the entity itself*'.[83] Subjection to capitalist social relations, while formed through the inversion internal to the fetish character, is also produced by the process of capital's realisation[84] and its reproductive form, 'money'.[85] Money constitutes a form that is immanent and external to the logic of capital. The subjection of human life to the reproduction of the value form (the life process of capital) retains its place as immanently external to the logic of capital. Human life persists as a transhistorical physiological entity in metabolistic relation to the historically constituted sensuous world through the

---

81   These are all Hegelian categories that articulate a self-reproducing subject, depicting a universality that is an 'objective immanent form'.
82   Marx 1992, p. 185.
83   Lange 2019, p. 158.
84   This takes place through circulation retroactively posited in production.
85   Characterised by its threefold nature as a measure of value, medium of circulation and universal equivalent.

mediation of labour. Human life and value form, with their distinct temporalities, are intrinsically connected through abstract labour (which is the substance of value). Abstract labour only appears as value for itself in the form of money, as the abstract objectification of human labour. Here, life itself *is* really subsumed to the logic of capital. Yet, as in the case of life in relation to the absolute idea in the *Logic*, life is subsumed in such a way that its otherness is necessary for the role it plays as a variable in the development of the concept. The very otherness of life is internal to the logic of capital, which is why its externality is immanent. These two elements (the life of capital and human life), through their ongoing practical mediation, create one another anew: People living in capitalist societies, and living within different relations therein, become different kinds of people with different historically determined needs.[86] While the movement of the value form shifts based on its encounter with the natural limits imposed by living elements, it simultaneously subsumes these limits into its own logic. Therefore, these elements remain fundamentally relational. Marx's ability to think their interrelation derives from his commitment to Hegel's logical method, where form is never developed in separation from matter.[87]

Analysis from the perspective of circulation and reproduction reveals a central contradiction of capital, where capitalist variables meet non-capitalist variables in dialectical co-existence. Importantly, 'non-capitalist' variables and forms (variables and forms developed with indifference to abstract labour) are here developed from the standpoint of capitalism and therefore do not come from 'pre-capitalist variables'; rather, they are non-capitalist variables produced by capitalist forms. This is a doubling of the fetish that functions to create its own opposite. Within circulation, human life and the sensuous world form natural limits to potential valorisation by existing with distinct temporalities. As a general equivalent, the money form imposes formal limits at the level of

---

86  For a detailed account of the historical specificity of need in Marx, see *The Theory of Need in Marx* (Heller 2018).

87  Interestingly, Canguilhem, in his essay 'Le concept et la vie', begins his detailed overview of the philosophical address of 'life' as it relates to 'concept' with the question of form. For Canguilhem, to think about the relation between life and concept, the first thing that one must consider is how life is the organisation of matter and, as such, the creator of forms rooted in the experience of a singular living being. Nonetheless, this presents a stark problem. How does conceptualisation of form give us access to the nature and meaning of life? Canguilhem reduces this question to its most basic articulation: How does intelligibility meet life? The relationship between intelligibility and life has long relied on the question of reproduction, as it is well known that reproduction plays a key role in the development of classification in Aristotle's formal account of matter (Canguilhem 1966, pp. 193–223).

the reproduction of the life function of capital as 'automatic subject'.[88] Money too must circulate without valorisation. Therefore, human life and the sensuous world, and the money form, containing temporalities distinct from capital's abstract form, relate through conceptual identity as mediums of reproduction.

The contradiction of capitalist and non-capitalist variables at the heart of the reproduction of capital – facilitated through its medium, money – gains clarity through analysis of credit money. Credit money exists as a form of fictitious value: as a condition that exceeds the substance of accumulated past abstract labour (which is retained and leveraged by the creditor or crediting body). While credit money is leveraged by past abstract labour and therefore depends on a fetishised value form as its condition, the credit money itself is rendered valid through interpersonal forms of domination that promise future production of value as substance, forged through a contract that takes the place of abstract labour as substance (the legitimising factor that makes it a real abstraction). This is why Marx called it fictitious: it lacks the value's defining substance because of its independence from production, which is practical and temporal.

## Conclusion

To facilitate reproduction, capital's realisation process relies on a supplement to the logic of capital as an internal logical necessity. Hence, understanding the dynamic of capital's reproduction, and of social reproduction, within Marx's critique requires a thorough understanding of the money form as an internal logical other within the logic of capital's realisation. Accordingly, to interpret social practices that reproduce capitalist form requires us to delineate these practices as indirectly value-producing activities, which remain formally other thereto. Marx's reproduction schemas within Volume II connect different variables and plural, seemingly fragmented and internally differentiated social practices within the process of capital's accumulation as a whole, articulated from the point of view of total social capital (or expanded reproduction, which includes the function of credit money an essential feature of the money form). Here, Marx's focus on circulation time places emphasis on how each variable contains its own temporality, for which synchronisation and limits are imposed by temporal variables. Natural limits cause temporal interruptions (we must sleep, and seasons change) accounting for non-capitalist time, or the time of

---

[88] Marx refers to capital as an automatic subject on p. 255 of *Capital* Volume I (1990).

life (as opposed to the time of capital), constituting necessary variables within the circulation process.[89]

This chapter has anticipated the next chapter, Chapter 5, 'Marx's Social Theory of Reproduction', where analysis turns to Hegelian philosophy to address fundamental questions of Marx's method. Chapter 5 draws upon Hegel's use of the term reproduction in the chapter on 'Life' in *The Science of Logic*. The logic of capital, or Marx's value theoretical exposition in *Capital*, is drawn from Hegel's *Logic*. It is here, therefore, that Marx's understanding of reproduction and its contradictory character in the movement of capital can be found. This retrieval correctively demonstrates that the reproduction of concrete life acquires determination in *Capital* as an ontologically distinct negation of the Hegelian logic of abstraction. Therefore – as other to capital's abstract form – concrete life is found to require a different philosophical system for its comprehension. This is not only a positivist account of organic life drawing on biology, but an epistemological position that incorporates aspects of human individuality derived from post-Kantian and Feuerbachian philosophical legacies. Drawing upon and reworking these philosophical tools, this final chapter will use these categories to construct a critique of political economy from the point of view of social reproduction.

The final chapter will argue that relations of freedom and domination are better understood through focus on 'reproduction' than through the category of 'labour', since the former encompasses the latter.[90] The merits of this orientation are particularly evident in our current conjuncture. Whilst wage labour remains necessary to capitalist reproduction and is in its own right an important site of struggle, financialisation has reconfigured its place to such an extent that it no longer clearly constitutes the central form of mediation between capital and the concrete, or human life and nature. This is exemplified by the status of unemployment, unpaid reproductive labour, debt and the destruction of the environment, to name only a few. This analysis – foregrounding reproduction as the locus of the primary contradiction between capital's abstractions and concrete life – is pursued with the purpose of locating how to undermine the dominance of capital's social form over concrete life. As David McNally has correctly assessed 'one of the tasks of the Left is to highlight this conflict – between life-values and capitalist imperatives ... ... in order to pose a socialist alternative that speaks directly and eloquently to the most vital needs of the oppressed'.[91]

---

89  Tomba 2009, pp. 135–7.
90  This is something generally incorporated into the position taken on labour in Marxist social analysis, even in the case of social reproduction theory, such as the work of Sylvia Federici, that theorises the reproduction of life from the point of view of value producing unpaid labour. See Federici 2012.
91  McNally 2009, p. 74.

CHAPTER 5

# Marx's Social Theory of Reproduction

## Introduction

Any thorough analysis of Marx's theory of capital's self-reproduction requires consideration of 'life'. Marx theorises capital's abstract capacity to self-reproduce through the notion of a 'life process'. This 'life process' describes the reproduction of capital's abstract social forms, implying another concept of reproduction: 'concrete reproduction'. 'Concrete reproduction' comprises the distinct processes of the reproduction of nature, matter and the lives of persons in capitalist societies. Thus, at the convergence of German idealism and the natural sciences, Marx advances two concepts of life that retain distinct processes of reproduction with different logics. As concrete on the one hand and abstractly capitalist on the other, these concepts of life together reproduce capital relations. The two are ontologically distinct, yet, in capitalism, they exist in permanent tension.

Concrete reproduction is determined by naturally limited life-making processes endowed with transhistorical capacity to reproduce without capital's abstractions. However, the process of concrete reproduction is also shaped by capital's self-reproducing forms. Capital's reproduction not only relies on human life to supply labour power, but also relies on ecological life, matter and mechanisms of circulation – all of which possess social and historical determinations inextricably linked to capital's abstractions. Marx's two concepts of 'life' grapple with the dynamic between abstract reproduction and concrete reproduction. In doing so, Marx overcomes the limitations of preconceived determinations of the 'concrete' – manifest in designations of which activities produce value – permitting an enriched understanding of practices and processes that exceed the scope of capital's abstractions. This is because capital's abstractions alone are unable to conceive much of the content with which they are mediated, such as biology, physiology and further affective qualities.

Recent literature on 'social reproduction' has brought Marx's analysis of capital's reproduction into renewed focus. For the most part, this writing has aspired to construct a 'unitary theory'[1] or a one that considers capital itself as

---

1 This term was inherited from Lise Vogel's book *Marxism and the Oppression of Women: Toward a Unitary Theory* (2013).

a system that reproduces gender and racial oppressions.[2] This is generally formulated in terms of capital's reliance on oppression for the reproduction of its necessary relations of accumulation: Oppressive relations bring about readily available labour power and therefore uphold the wage relation. Oppressive social practices, such as racial hierarchies, reproduce specific compositions of labour power, and they are also formed, sustained or curtailed depending on the availability of wages – or the ability of labourers to claim them. Aligned with many of the conclusions reached in the social reproduction literature, the orientation adopted here seeks to deepen the analysis by considering the logical role of the reproduction of capital as well as the place of non-capitalist forms and practices within the movement of capital's abstractions. Marx owes his systematic method to Hegel in making the object of knowledge in *Capital* a social totality. In doing so, Marx proceeded not historically but logically, with capital expounded as an abstract universal that posits its own conditions. The aim of this chapter is to better understand capital's self-movement. To do this, it draws on Hegel as a fundamental source for reconstructing Marx's method. If done correctly, a reconstruction can reveal points of contestation in logical externalities within capitalist totality. These logical externalities are both necessary conditions for accumulation and presuppositions posited by capital.

In much of the Marxist canon, labour, in its concrete manifestation, has been thought in relation to capital's abstractions and has correspondingly been interpreted in terms of the movement of abstract value forms, including the money form and the commodity form. Social reproduction theory has often departed from this for an evaluation of the way unpaid labour – or toil – reproduces the labour power as a commodity. My analysis aims to engage both sides of this divergence by contributing the proposition that social and capital reproduction are better understood with a reorientation to the reproduction of life, human and natural, analysed through its mediation by capital's abstractions.

The Marxist canon has predominantly focused upon the wage relation. Consequently, the interpretation of the reproduction of human life, and of nature more generally, has largely been addressed secondarily to understand wage labour. Social reproduction theory has sought to address this bias. This has been done through an insistence that Marx's critique of capital is not based on the separation between economic and non-economic spheres, and therefore that both sides are central to the reproduction of the social relation. Unpaid labour and practices of care that reproduce life are argued to be just as essential to capital as the wage because unpaid labour sustains the reproduction of

---

2  See Arruzza 2014, 2015a, 2015b; Gonzalez and Neton 2014, pp. 149–74; Vishmidt 2012.

paid labour. Meg Luxton expresses this in her insistence that 'the production of goods and services and the production of life are part of the same integrated process'.[3] This is echoed by Susan Ferguson when she claims that 'the social reproduction of labour cannot be separated from the social reproduction of life',[4] and by Nancy Fraser when she states that '"the crisis of care" is best interpreted as a more or less acute expression of the social-reproductive contradictions of financialised capitalism'.[5] Hence, social reproduction theory looks at capital from the perspective of the co-production of the economic and the non-economic as an indispensable point of analysis. However, in social reproduction theory thus far, the coexistence of the economic and non-economic is largely understood through the tension between the reproduction of human life and wage labour, and not between human life and capital's process of reproduction.[6] Analyses that do the former, while accounting for the reproduction of the variable of wage labour, fail to account for the reproduction of capital as a social form. Hence, much of social reproduction theory – whilst examining the *reproduction of labourers* – has lacked systematicity in its critique of capital, rendering it inadequate to determine the relationship between the respective registers that it aims to grasp as interlinked: concrete and abstract reproduction.

The separation of social reproduction from a logical interpretation of the reproduction of capital's abstractions – present in both traditional value-theoretical Marxism and social reproduction theory – begets a labour-centric account of 'social reproduction'. When analysis considers social reproduction merely as the mechanism that reproduces labour power – and not as a concrete manifestation internal to capital's abstract movement to self-reproduce – it fails to include an analysis of the social reproduction of those who are excluded from wage labour. Moreover, the precise role of finance[7] and ecology in capital's reproduction are obscured. Finance (and financialisation) entail a monetary process of circulation and reproduction of capital's abstractions,[8] while ecology provides the necessary material for capital's sustenance. The two indirectly

---

3   Luxton 2006, p. 36.
4   Ferguson 2017, p. 129.
5   Fraser 2017, pp. 21–2.
6   Nancy Fraser's work is one such exception. However, her analysis of financialisation is embedded in a Polanyian idiom that misses aspects of Marx's development of the relationship between life and form.
7   Where credit money, or money as money, is an immanent other to capital's abstraction that makes money the medium of circulation and therefore the medium of the reproduction of capital.
8   This process was considered in detail in Chapter 2, 'Money Form'.

affect the capital-wage relation and are thus only partially comprehendible from the perspective of wage labour's reproduction.

Abstract logic and concrete social experience are not identical within capitalism. They possess a dialectical relationship. While they might exist in opposition, their identities are co-mediated. In this regard, a logical approach to thinking concrete reproduction ought to address how concrete experience of capitalist societies – including relations of interpersonal forms of domination – relate to capital's abstractions. Without a logical assessment of what reproduction is in capitalist social relations, a lack of systematicity will inevitably result, engendering a vague application of Marxian concepts (such as use value, value, abstract labour etc.). Such concepts, when thought in their theoretical-logical specificity, have significant political repercussions, particularly for how actors might conceptualise the difference between capitalist relations and practices and non-capitalist relations and practices. While often used to sustain gross inequality necessary for wage relations in capital, 'non-capitalist' practices are disciplined by interpersonal forms of domination, such as colonialism, imperialism, law and ideology (as was addressed in detail in section 3.2, 'Personal and Impersonal Forms of Domination'). Therefore, a strong conceptual differentiation of capitalist and non-capitalist relations, combined with an analysis of their logical interrelations, might furnish the tools to identify both emancipatory possibilities and to more precisely analyse historical developments, determining points at which non-capitalist practices contradict or bolster capitalism's logic at large.

The immanent and external nature of money as a reproductive form of capital's abstract movement of value shows that, logically, capital requires what is ontologically other to it as a medium of its reproduction. Drawing from this insight that demands logical assessment, in this chapter I claim that Marx's theory of the reproduction of capital logically requires capital's other. That is, a theory of reproduction needs to understand the logical role of concrete and historical life-making processes as external and immanent to capital's abstractions.[9] Here, a Marxian theory of reproduction will be examined by drawing

---

9 This framework gives us tools to engage with debates where the theorisation of the relation between gender and capital is often deployed as either a unitary theory or the intersection of different institutions of domination. Instead, this framework shows that the relation between the concrete and the abstract are dialectical, and while in line with Meiksins Wood's argument of the indifferent nature of capital's abstractions to racial and gender-based oppressions, it departs from Wood in the claim that, dialectically, these oppressions have become necessary and immanent in practice as social forms that reproduce the capital relationship. Although capital might not need gender oppression, it does require gross social inequality to be internal to the class of the wage labourer. Gender and race inequality has in practice

upon Marx's articulation of the value form. Doing so affords perspective on how the current state of the reproduction of life and nature (as outlined by antiracist, feminist and environmental Marxisms) reflects dynamics internal to capital as a social form. In other words, life and nature will be grasped as 'structural' for capitalism. The upshot of this orientation will be a renewed call to foreground the critique of capital in theories of social emancipation and ecological reclamation.

In the analyses so far in this text, a theoretical framework from which to examine the relationship between capital and its outside from the perspective of value has been established. In previous chapters, this was done by establishing the immanently external nature of the money form and the distinction between impersonal and personal forms of domination. Doing so enabled examinations of capital's relationship to 'life', both that of humanity (social reproduction) and that of nature. Human life and nature are two variables to capitalist totality, upon which capital relies but does not directly produce. In *Capital*, the circulation of value is parasitically dependent on its concrete other as a means of reproduction, requiring human life and nature, to sustain itself, and to absorb waste products. For Marx, human life is not opposed to nature but is transhistorically dependent on it, existing in a metabolic relation thereto. In Chapter 5 of *Capital* Volume 1, 'Labour Process', Marx defines labour as the following: 'a process between man and nature, a process by which man, through his own actions, mediates, regulates, and controls the metabolism between himself and nature'.[10] At the same time, this metabolic relationship is altered by capital's abstractions. Life-making activities, as reproductive social forms, cannot be neatly isolated from capitalist activities; however, they retain a transhistorical character, rendering them potentially non-capitalist. Capital determines the way in which life-making activities are undertaken as modes of reproduction internal to capital's own self-movement. And, in dialectical relation to capital's abstractions, history has concretely determined what constitutes 'nature' through the ongoing process of destruction and regrowth.

Capitalism largely treats nature and human life as self-replenishing and readily available, to be used to produce value without functioning *a priori* as capital but as material outside of capital. They are often extracted as use values

---

provided for that requirement, rendering them necessary for the historical process. Interpersonal forms of domination have thus been immanently reproduced in capitalist relations. Although interpersonal domination is external to the specificity of capital's abstraction, these forms of domination are concretely reproduced.

10   Marx 1990, p. 283.

and not compensated for as exchange values. In this way, capital's reproduction logically disavows its own means of reproduction, which is a requirement for the means of reproduction to remain external to capital's abstract system of values. This becomes clear when we look to the disavowal of the social activity that goes into the upkeep of the labourer and the lives of people in capitalist society more generally: Social reproduction is necessarily external to value production, which is why it can be a reproductive agent of the capital relation.[11]

The logic of the social reproduction of capital will always rely on the disavowal of what is other to it. Irrespective of changes occurring at the level of social rights, capital will nevertheless continue to degrade the mechanisms that sustain the extraction of profit. If a given society, for example, achieves aspects of gender equality in social reproduction, the exploitation of social reproduction will shift to exacerbate another form that will likely carry a racialised and gendered character. Social reproduction, or the reproduction of labour power, is racialised and gendered as a historical reality, and its gendered nature has thus been, in practice, necessary to capital. Capital, therefore, should not be understood as a practical reality without its gendered or racialised composition: logically, capital's development becomes internal to capital itself though the mediation of the concrete by the abstract. However, this does not mean that capital cannot function otherwise: Capital's abstractions, as abstractions, possess an indifference to the concrete (we do not get a distinction between use value and exchange value without an indifference to the concrete on the side of the mediation of the abstraction). Hence, interpersonal forms of domination can develop and regress within the framework of a capitalist mode of production. Evidently, it is possible to have more and less egalitarian forms of capitalist relations.

Capital extracts value from where value did not yet exist, such as unpaid labour and materials from nature.[12] These inputs are taken as cheaply as possible while, as self-expanding value, capital accumulates infinitely. Therefore, capital's immanent relationship to its other is also what renders it the sociohistorical driver of climate change.[13] Capital parasitically draws on life and nature as a necessary outside, or as inputs that provide material for capital

---

11   Chapter 2, 'Money Form', has shown the money form to occupy a logically comparable position in Marx's thought.
12   Marx directly makes the claim in *Critique of the Gotha Programme* that 'Labour is *not the source* of all wealth. Nature is just as much the source of use-values (and surely these are what make up material wealth!) as labour' (2019, p. 1025).
13   Fraser 2021, p. 127.

to consume. In both cases, capital does not systematically replenish either one, leaving labour to be replenished by social reproduction, a form of social renewal that utilises practices of care distributed through gendered and racial hierarchies put into place within capital's self-development. Nature is devalued as much as possible – if not extracted without any exchange – and taken without reparation of damages done. This is a dynamic that Fraser rightfully terms 'discounted ecological-reproduction costs'. Fraser observes that 'not "just" raw materials, energy, and transport, but also labour as wages fall with the cost of living when capital wrests food from nature on the cheap'.[14] Capital's reliance on unpaid labour (as both unpaid social reproduction and wage exploitation) and environmental extraction are necessary for the possibility of profit. Fraser explains,

> In every case, capitalists appropriate the savings from cheap inputs in the form of profit, while passing the environmental costs to those who must live with – and die from – the fallout, including future generations. More than a relation to labour, then, capital is also a relation to nature – a predatory, extractive relation, which consumes ever more biophysical wealth in order to pile up ever more 'value', while disavowing ecological 'externalities'.[15]

This chapter looks to expand and deepen Fraser's argument. In doing so, it returns to the Hegelian roots of Marx's critique of political economy to develop a robust interpretation of capital's reproduction, which is driven by its relationship to the reproduction of social life and nature.

The category of reproduction in Hegel's *The Science of Logic* – found within the sections on Life in the second volume, 'The Science of Subjective Logic or the Doctrine of the Concept' – anticipates Marx's critical exposition of capital's 'life process' [*lebensprozesses*]. For Marx, as for Hegel, reproduction always entails a self-moving, living element, which may occur at the level of an abstraction – as in Hegel's absolute spirit and in Marx's capital.

As I will discuss in greater detail, capital possesses a 'life process' because it self-reproduces as an autonomous subject. In Marx, capital is described through the term 'life' because it self-reproduces. Therefore, my interpretation of life and the living requires an interpretation of reproduction, while reproduction, in turn, requires an interpretation of life (since reproduction is a life-

---

14  Fraser 2021, p. 100.
15  Fraser 2021, pp. 100–1.

making activity). By analysing reproduction as a mechanism internal to capital's self-reproduction, or life-like character as 'subject of the process' (which is the subject of the movement of value throughout the value forms), I locate the central contradiction of capital in reproduction, where exploitation, realisation and accumulation are sustained.[16] This is a contradictory process because, throughout the reproduction of capital, resistant heterogeneities of 'immanent externalities' remain necessary for the continuation of capitalist accumulation. In this regard, the reproduction of social capital as totality, induced by capitalism itself, is irreducible to the movement of value as form only in so much as the value form is an abstraction that parasitically mediates what is other to it.

To develop an interpretation of this contradiction in reproduction, this chapter grapples with a difficulty intrinsic to Marx. In his methodological reliance upon Hegel's *The Science of Logic* for his exposition of the value form, Marx's thought remains Hegelian. However, this Hegelianism remains only partial. Marx employed two different concepts of life. One concept develops from the Hegelian dialectics of the *idea*; it is completely fetishised to elucidate the life of capital – which, as a fetishised form of autonomy, is only ever life-*like*. By contrast, the other concept of life is physiological, rooted in natural science. These two concepts represent two conflicting and mutually positing processes: capital's self-expansion on the one hand and the reproduction of human life and nature on the other.

Marx's physiological concept casts 'life' as an ontological prerequisite for capital as a process, which, strictly speaking, remains un-Hegelian. This makes it possible to identify a bifurcation: in Hegel, an ontological prerequisite is completely internal to the concept and therefore becomes its own form of transition from external to internal. This leads not to a 'resilient' contradiction but to a reconciliation. In contrast, in Marx, the ontological prerequisite of concrete life remains conflictual and produces permanent interruptions and obstacles to the concept-subject-capital, which has acquired an autonomous life. This bifurcation occurs because the life process of capital is entirely fetish-

---

16  This is essentially a contradiction between material content, or life in metabolic relation to nature, and abstract form. This is in contrast to more traditional approaches that locate labour as the central point of tension in capital. Alienated labour, however, is a capitalist activity. Although labour power retains a contradictory structure, as a commodity that has a use value and an exchange value, wage labour is not a contradictory practice: it is a capitalistic practice. Instead, the essential practical contradiction – between non-capitalist practice and the processes sustaining capital's development – occurs in the reproduction of capital. Wage labour is immanent to this process.

ised and abstract, while the life of those living within capitalist social relations remain concrete. In the dialectically reunified category of 'living labour', which contains both the abstract and the concrete, and therefore two concepts of life, concrete natural life[17] and the life of capital exist in permanent tension (without reconciliation). Concrete natural life and the life of capital meet in living labour. Therein, the two concepts of 'life' form an antagonism whereby concrete life is completely internal to the realm of social reproduction, or the reproduction of human life in metabolic relation to nature.

The 'life' that capital parasitically lives off is a positive concept of life. Its distinction from the Hegelian concept of life registers an ontological difference that highlights the parasitic nature of capital and the corresponding impossibility of cohabitation between capital's process and the replenishment of life, be it social life or nature. This disjunction is illuminating in that it helps us to think through the ecological crisis, emphasising the impossibility of the affirmation and renewal of nature within the context of capital. Likewise, the two different concepts of life highlight the impossibility of the affirmation of human reproductive activity as self-determining within a capitalist mode of production.

The role of Marx's positive sense of life, which draws on a combination of natural science, in particular physiology – and the critique of Feuerbachian humanism – is interpreted in its function as a negation of the abstraction of capital's self-movement or life. How this sense of life acquires determination, not only in Marx's own thought but also in the concrete present, would require a detailed address of current research in natural science and physiology. The analysis in this chapter is oriented not to the ontology of this sense of life (which would require its own meaningful research project), but to how this second sense of life simultaneously works logically, within the movement of capital, as a negation of capital and as its means of reproduction.

In this chapter, I will first locate this project within the positions of relevant commentators working on the relationship between Marx's value form analysis and Hegel's *The Science of Logic*. I will then provide an interpretation of the passages in *The Science of Logic* that refer to reproduction. Finally, I will employ a comparison of aspects of *The Science of Logic* and *Capital* with attention to their respective theories of life and 'life process' to develop an account of reproduction derived from Marx's two concepts of life.

---

17  'Individual life' as both an existential regime of temporality and an organic physiological aspect of nature.

## 5.1    Capital's Life Process

Hegel's *Logic* constitutes the key resource for comprehending Marx's method within the critique of political economy. The object of knowledge in *Capital* is an abstract universal whole, or capital in its totality, which derives its theoretical systematicity from *The Science of Logic*, where Hegel's concept of 'spirit' is structurally adopted by Marx in identity (and non-identity) with his concept of capital. Because Marx's presentation across all three volumes of *Capital* is partial, the universal concept of capital at times requires methodological reconstruction with the help of Hegel. Although his project was unfinished (Volume I was the only part of *Capital* published in his lifetime), it also simply would not have been possible for Marx to present all the content of the universal concept of capital and its self-movement, since the movement of capital logically exceeded the historical epoch in which he was writing.

Hegel's writings are the source of the concept Marx employs to comprehend capital as an abstract universal whole, or as a universal subject with a 'life process'. Therefore, the dialectical form of presentation in *Capital* is enriched by going beyond Marx's own formulations.[18] Readers can do so by employing Hegel when gaps surface in Marx's elaboration of capital's development, as this chapter does when considering Marx's elaboration of capital's reproduction. The mediation of 'reproduction' within the life process of capital is an element of Marx's critique that has remained partially exposed. It therefore benefits from a reading of Hegelian logical exposition to establish the concept of capital. Because Marx's presentation often elides methodological systematicity, comprehension of his method requires rethinking his project at a different level of abstraction to that evident in *Capital*.

To analyse Marx's theory of reproduction, it is instructive to consider the related concept of 'life'. To do so, this section examines Marx's exposition of the abstract 'life of capital' and the implications it has for capital as a concept by drawing on theories of capital as subject. In doing this, the section locates Marx's use of such terminology in *Capital*. Attention to Marx's terminology reveals that the abstract concept of 'the life of capital' is dialectically dependent on a positive concept that denotes the 'life' of humans and nature. This establishes the theoretical basis upon which the rest of the chapter proceeds,

---

18    Additionally, reflection on the 'Young Hegelian' 'concept' manifest in the work of David Friedrich Strauss, Bruno Bauer, and Ludwig Feuerbach can illuminate Marx's method.

addressing the intersubjective logic at work between capital and its other, with Hegel's exposition of reproduction applied to capital and concrete reproduction respectively.

A key aspect of Marx's abstract exposition of the life of capital is a theory of capital as subject. The identification of capital as a 'subject' entails that capital comprises an element of freedom: that which drives 'the process'. Logically, Marx established capital to be the subject of the process because capital's abstractions automatically self-reproduce with the purpose of accumulating more capital. Value moves between commodities and, in doing so, accumulates capital at the point of exchange. However, the logic of capital does not exist in an a-historical, objective, logical vacuum that transcends material interference or the agency of people living in capitalist societies. As a self-moving logical process, capital requires what is other to it as a means of its own reproduction. Capital's abstractions are shaped by that which is external to it on account of limitations imposed by the concrete.

There is no dualism between the economic and the social. There are no laws of motion within capitalism that when formalised as a logic become antithetical to the historicisation of social development. The persistence of relations that are not specific to capitalism account for social changes within capitalism, which is evident when we examine the 'reproduction' variable. Reproduction is the primary site of contradiction between concrete practices of individual agency – insofar as the individual is a historically constituted 'personality' – and capital's logical self-movement. Marx demonstrates that both human life and the life of capital have agency. Correspondingly, they require consideration of an intersubjective tension: both on the side of the human subject and the subject of the process that is capital (two sides that are co-determining and wholly relational). Hence, freedom determines the grounding for both Marx's anthropology and his understanding of capital. These two sides come into contradiction due to competing interests behind their respective free movement. In order to analyse this contradiction, this chapter looks to the logical dynamic of capital's subjective life process and subsequently interrogates the ontologically distinct concept of human life, which is equipped with its own formulation of subjecthood.

While Marx conceives of the human subject in terms of its own distinct anthropology, capital as subject is thought, drawing upon a Hegelian method, in terms of substance and form. This method entails a series of mediations. Marx's development of the concept of the subject- concept-capital is hence determined through a method and mode of presentation that has its basis in Hegel's *Logic*.[19] The subject-object inversion at the heart of the fetish character

---

19   This analysis does not follow a homology theory of the influence of Hegel's *The Science*

of capitalist society is a material realisation of a process of reflection determinations that determine the subject through the its dialectical relationship to the object. This dialectical relationship is founded on the opposition of being and essence (which are separate sides that are at the same time inseparable). The reflection determinations that arise are a series of relations between categories based on a process that goes through a series of mediations ('qualitative leaps') and ultimately arrives at the developed concept of the subject. The reflection determinations that present the process of the emergence of the subject are drawn upon in Marxian readings of Hegel's *The Science of Logic* to represent the culmination of the objective logic (as found in books I and II). For Marx, the process in which these reflection determinations unfold is the process of valorisation, where mediations occur between the different value forms (commodity, money, labour etc.), eventually developing into capital. The movement is directed towards the accumulation of capital, which is why capital takes the place of the subject for Marx.

Within the movement of value through its different forms, the movement of being contains value, which is 'essence'. Reflection, is a moment in the *Logic* that is determined by the continual dualism between being and essence. Thought splits [*Gedanke*] into these two sides – sides that are at once inseparable and separate The two sides 'face each other' in reflection as both identical and non-identical. In *Capital*, we find reflection determinations (identical and non-identical figures in co-existence) in the relationship between value and price, surplus value and profit, and the value of socialised labour and the wage. Here, profit is the being of 'surplus value' and price is the being of value, while surplus value and value are the essence, etc. As is the case in *The Science of Logic*, in *Capital*, distinguishing between being and essence, where both are identical and non-identical, is difficult, as the differences therebetween are determined by the different levels of abstraction required for comprehension. For example, profit is calculated and appears concretely, while surplus value is an abstraction in thought alone.[20]

Marx uses Hegel's philosophy of essence to show how, in capital, what exists in the immediate has a 'reflexive existence', where the objective world really

---

   *of Logic* on *Capital*, as in the case of Arthur, but rather, following Lange, Tombazos and Heinrich among others, it maintains that Marx utilises Hegel's *method* and that therefore these categories are not homologous to Hegel's. Nevertheless, categories in Hegel's *Logic* can help illuminate Marx's critique, and certain Hegelian categories are utilised by Marx in a reified form. Marx has a much more complicated presentation than that of mere homology, where the subjective is not always endowed with Hegelian inflection (i.e. an Aristotelian influence underpins much of his thought).

20  Tombazos 2015, p. 72.

exists as it is thought (and not only its being), which is at the same time true of the object as being. Here, the object of thinking becomes more 'true' than its corresponding object.[21] The object of thinking, such as surplus value in this example, requires additional mediation than a reflexive dualism and is what Hegel means by the 'Concept' [*der Begriff*], which is speculative. Here, clarification of the role of the subject in the process of capital's accumulation requires focus on Volume II, Book III of *The Science of Logic*, 'The Subjective Logic'. In the section 'The Subjective Logic', the conceptual relations necessary to think 'the subject' are grasped at the meta-level of the concept of the concept. It is to this that I now turn.

The concept – the unity of being and essence – is immanent to the two-fold relations of being and essence in their separate manifestations. This takes place within the movement of value in individual capital, as the foundation of the value forms and as what Mark E. Meaney rightfully refers to as the very 'organising principle'[22] of their movement. Once capital is described from the point of view of total social capital, or the unity of production and circulation, the two-fold relations outlined by Hegel are effectively superseded as concept: capital becomes a self-causing whole or a totality of moments that are inseparable from one another. Yet this whole has always already presented itself as its own organising principle. In this way, as Meaney astutely observes,

> Capital as a unity of production and circulation is that "organic unity" which is the ground of the entire development. In its "immediate being," capital is commodity and monetary circulation.[23]

As an 'organic unity', capital supplants the reflexive dualisms internal to the process of accumulation and becomes a speculative concept. As Stavros Tombazos points out, going beyond the reflective dualism is the action of the Concept/Notion [*der Begriff*].[24] This is the last moment of the three sections of *The Science of Logic*. When applied to *Capital*, the transcendence of reflective determinations enables a reading of capitalist economy enriched by both process and practice, which together create a social totality.

To think capital as concept, as Meaney and Tombazos have done, requires exploration of Hegel's influence on *Capital*. But it does not necessitate a direct

---

21   Tombazos 2015, p. 74.
22   Meaney 2002, p. 114.
23   Meaney 2002, p. 113.
24   Tombazos 2015, p. 73.

mapping of the movement of being to essence to concept.[25] Instead, Marx's methodological application of the *Logic* corresponds to *Capital* with more consistency only in the third book, *The Doctrine of the Concept*. Marx does this to better understand the function of the subject in *Capital* and, accordingly, how capital as subject influences the subjection of concrete life to its form. In this sense, the work that the subject/concept/notion performs in *Capital* has results that are inconsistent with their conceptual roots in the three parts of *The Science of Logic*. These conceptual roots are combined and articulated differently by Marx on account of their practical subjection to the concept of capital. As the result of the process, the 'concept' is a category that engenders an understanding of the economy as a social totality with complex internal contradictions. Marx derives his understanding of the 'economic world' from the requirements of the concept or notion. To bring conceptual clarity to Marx's interpretation, it is instructive to integrate Hegel's terminology with Marx's terminology.

When drawing on Hegel's *The Science of Logic* to understand the method of *Capital*, it is necessary to keep in mind that Hegel's *Logic*, although called *The Science of Logic*, is a book about metaphysics. Upon first glance, it is questionable as to why a book of logic might deploy categories such as 'being', 'essence' and even 'life'. In Hegel, these categories recur because *The Science of Logic* is less a meditation on logic as a discourse than it is a development of epistemology and ontology. Hegel's logic *is* his metaphysics, albeit a metaphysics conditioned by a logical method formulated at the level of epistemological questioning. Contradictions are overcome by this epistemological questioning, which leads to double negations such as the question, 'what is the concept of the concept?' Such questions epistemologically demand a logical derivation. As Christopher J. Arthur shows, the truth of this logical method is 'meant ontologically as much as logically' and 'the coherence of the logic is at the same time the coherence of reality'.[26] For Hegel, metaphysics must have a logical method because, according to his work, being is relational. And, in being relational, 'being' is also temporal. Therefore, 'what is' comes into being by becoming in space and time relationally to its other. This relation results in a sublimation of one category into another, creating a more fundamental category. For Hegel, knowledge of a category cannot be grasped in its immediacy. Rather, it requires processual movement in order to actual-

---

25   This is in contrast to Christopher J. Arthur's project put forward in his book *The New Dialectic and Marx's Capital*, which maps Hegel's *The Science of Logic* onto Marx's *Capital* to construct a conceptual homology.
26   Arthur 2002, p. 84.

ise its concept. This is why Marx draws on Hegel. Marx, whose primary concern is to reveal the nature of capital, sees that what is in existence exists as a social relation. Therefore, what exists came into being through a set of social practices that are co-determining. A social relation is something that exists in practice and therefore cannot exist *a priori*: it is a result of a process.

Attention to the function of Hegelian logic in *Capital* reveals the role of 'life' in the text. There, life operates as an ontological precondition that acquires two different characters, referring on the one hand to human life and nature, and on the other to the life of capital. Life retains a double character throughout all three volumes of *Capital*, but it receives greatest focus in Volume II. Volume II exposits capital's circulation and, correspondingly, reproduction, through the application of a functional concept of the metabolism of forms [*formwechsel*]. Marx understands the self-reproduction of capital's abstract form as the reproduction of capital as 'subject' of the process, entailing capital's endowment with life-like features. Marx uses phrases to demark his theory of capital's 'life process' in all three volumes of *Capital* as well as in the *Grundrisse*.[27] This includes two decisive passages from Volume I.[28] In Chapter 10, 'The Working Day', Marx states:

> But capital has one single life impulse, the tendency to create value and surplus-value, to make its constant factor, the means of production,

---

27  While not a text that this book relies on due to complex continuities and discontinuities between it and the three volumes of *Capital*, relevant passages from the *Grundrisse* include the following: 'circulation is an inescapable condition for capital, a condition posited by its own nature, since circulation is the passing of capital through the various conceptually determined moments of its necessary metamorphosis – its life process' (Marx 1973, p. 581; for more references to capital's life process in the *Grundrisse*, see pages 543, 560–1, 569, 598 and 797).

28  Although 'life' is found in Chapter 4 of the edition of *Capital* Volume I with which we are working, the term does not appear in the same instance in the original German. The use of the term life is more restrictive in the original German, indicating that its use is to describe something more philosophically specific. The passage where the translation includes life while the German does not is the following: '[value] is constantly changing from one form into the other [commodity and money], without becoming lost in this movement; it thus becomes transformed into an automatic subject. If we pin down the specific forms of appearance assumed in turn by self-valorising value in the course of its life, we reach the following elucidation: capital is money, capital is commodities. In truth, however, value is here the subject of a process in which, while constantly assuming the form in turn of money and commodities, it changes its own magnitude, throws off surplus-value from itself considered as original value, and thus valorises itself independently' (Marx 1990, p. 255).

absorb the greatest possible amount of surplus labour. Capital is dead labour, that, vampire-like, only lives by sucking living labour, and lives the more, the more labour it sucks.[29]

And in Chapter 11, 'The Rate and Mass of Surplus Value':

Instead of being consumed by him as material elements of his productive activity, they consume him as the ferment necessary to their own life-process, and the life-process of capital consists solely in its own motion as self-valorizing value.[30]

In *Capital* Volume II, Marx refers to life more frequently, writing:

In the life of capital, the individual circuit forms only a section that is constantly repeated ...[31]

And further:[32]

... the periodicity in the capital's life-process, or, if you like, the time required for the renewal and repetition of the valorisation and production process of the same capital value.[33]

The terminology is carried over into Volume III. In the first chapter, Marx writes:

In Volume I we investigated the phenomena exhibited by the *process of capitalist production*, taken by itself, i.e. the immediate production pro-

---

29   Marx 1990, p. 342. A more direct translation I have made is: 'Capital has its own life instinct, the instinct of value itself'. The original German reads: 'Der Kapitalist hat seine eigne Ansicht über diess ultima Thule, die nothwendige Schranke des Arbeitstags. Als Kapitalist ist er nur personifizirtes Kapital. Seine Seele ist die Kapitalseele. Das Kapital hat aber einen einzigen Lebenstrieb, den Trieb, sich zu verwerthen, Mehrwerth zu schaffen, mit seinem Consta uten Theil, den Produktionsmitteln, die grösst- mögliche Masse Mehrarbeit einzusaugen, Das Kapital ist verstorbene Arbeit, die sich nur vampyrmässig belebt durch Einsaugung lebendiger Arbeit und um so mehr lebt, je mehr sie davon einsaugt. Die Zeit, während deren der Arbeiter arbeitet, ist die Zeit, während deren der Kapitalist die von ihm gekaufte Arbeitskraft consumirts. Consumirt der Arbeiter seine disponible Zeit für sich selbst, so bestiehlt er den Kapitalisten' (Marx 1867, p. 200).
30   Marx 1990, p. 425. This passage appears in the original German, pp. 289–90.
31   Marx 1992, p. 235.
32   Other passages from *Capital* Volume II can be found on pages 248, 427 and 508.
33   Marx 1992, p. 236.

cess, in which connection all secondary influences external to this process were left out of account. But this immediate production process does not exhaust the life cycle of capital. In the world as it actually is, it is supplemented by the *process of circulation*, and this formed our object of investigation in the second volume. Here we showed, particularly in Part Three, where we considered the circulation process as it mediates the process of social reproduction, that the capitalist production process, taken as a whole, is a unity of the production and circulation processes.[34]

Marx goes on to state that,

... capital runs through the cycle of its transformations, and finally it steps as it were from its inner organic life into its external ...[35]

These descriptions of capital as a living process rely on Marx's commitment to thinking the subject character of capital as self-reproducing: as a process that occurs within the circulation of capital where capital reproduces itself – as does social life. Marx reminds readers that *the circulation process mediates the process of reproduction*. While there are instances in *Capital* in which capital's abstractions contain a vital character that is patently metaphorical, such as in references to the circulation of money as being akin to that of blood,[36] the logically Hegelian role of life attributed to capital as self-reproducing – or, put differently, the logical movement of value as a metabolism of forms [*formwechsel*] – is not metaphorical.

Human life, in contrast to the life of capital, has an intrinsically interconnected double role: it plays the role of reproducing itself for itself and the role of reproducing itself as a variable for the accumulation of capital. These two roles have competing interests. Social conditions produce and uphold these abstract forms and are required for capital's valorisation. Yet the abstract forms contradict the life interests of persons in capitalist society. This is not just the case for the working class: Capital's valorisation degrades nature in a process injurious to the whole of society. While human life provides capital with substance through the abstract labour generated in exchange, it also provides the con-

---

34  Marx 1991, p. 117.
35  Marx 1991, p. 135.
36  For example, Marx refers to life in a metaphorical sense when he states in the *Grundrisse*: 'the human body, as with capital, the different elements are not exchanged at the same rate of reproduction, blood renews itself more rapidly than muscle, muscle than bone, which in this respect may be regarded as fixed capital' p. 670.

crete content of the bearers of the different value forms within the circulation process. Marx's articulation of abstract labour, the substance of value, connects human life to the life-like character of capital – but only in its abstract homogenised form as abstract labour. As I have discussed in Chapter 3, living labour can become abstract labour only due to the fetish character of the value form, which inverts the labour variable from subjectivity to objectivity through the abstraction of the substance of value (abstract labour). Value is a social relation that really is treated as a thing. While abstract labour is the substance of value due to its rootedness in human life or labouring activity, value is not the result of labour but of abstract labour. This means that there is no abstract labour in the production process; it only comes into being in the exchange process. In the concrete mechanism of the production process, human life must be concretely self-sustaining and the ability for one to do this is often undermined by capital's valorisation.

Addressing the individual commodity, Hegelian Marxisms have typically prioritised the first two chapters of the *Logic*, linking *Capital* with 'The Doctrine of Being' and 'The Doctrine of Essence'. By contrast, the argument presented here restricts its exploration to the third part of Hegel's *Science of Logic*, 'The Doctrine of the Concept', and does not look at the first two books directly. Because grounded in the exposition of the money form and its role as medium of circulation in expanded reproduction, this interpretation conceptualises capital from the point of view of the more fundamental categories that are developed after the logic of exchange has already become self-subsistent. Capitalist social relations are considered in terms of expanded reproduction – and therefore generalised commodification or subsumption – and not individual finite commodities. Money and commodities are finite objects only when looked at independently from capital. However, as an independent expression of value (money as money) in capitalist social relations, money cannot be thought without capital. What this also means is that commodity and money are finite objects only when they are looked at independently from reproduction. Therefore, without a method adequate to generalised commodification, one would lack the tools to address the reproduction of capital. This is emphasised by Tombazos when he observes that commodities and money as finite objects 'possess an aspect of "untruth" and "finitude" in the sense that they lack the moment of reproduction'.[37] Logically, a methodological untruth related to the finitude of the appearance of things is present at the level of presentation of the first two books of *The Science of Logic*. This is because categories first must

---

37  Tombazos 2015, p. 150.

appear before they can engage in the mediations that give them their truth. As Elena Lange has explained, the first two books are not lacking because of their 'untruth', but instead represent a 'semantic and pragmatic "cleft"' at the beginning of the presentation. This 'semantic and pragmatic "cleft"' internal to the least developed categories at the beginning of Hegel's exposition (being and nothingness) is a necessary moment in the dialectical development of the concept. With the exception of the final logical determination, each logical category contains this pragmatic discrepancy [*semantisch-pragmatische Diskrepanz*]. What this entails is that the meaning, rendered explicit by each logical category, does not express what is implicitly presupposed for its meaning.

This semantic presupposition is a necessary aspect of Hegel's logical method and is utilised by Marx. In Marx, too, incomplete categories from the beginning contain their opposites, which are the complete categories. This is evident in the categories deployed at the beginning of *Capital*. For example, the category of abstract labour, as the substance of value, requires the entire system of derivations as its presupposition. As a dialectical mode of exposition, the semantic presupposition of the category permits a critique of positivistic definitions, substituting them with a mobile and anti-dogmatic, historical form of thinking.

Hegelian Marxist polemics, such as Lange's, have emphasised the semantic presupposition of the category in order to develop a critique of Arthur's *New Dialectic*, in which Arthur rejects that it is methodologically necessary to begin *Capital* with the labour theory of value (as is the case in Marx's exposition). Arthur claims that labour as a starting point is a premature beginning that wrongly commits the exposition to simple commodity production.[38] However, as Lange points out, this claim departs from the dialectical character of Marx's method and therefore negates the significance of influence of *The Science of Logic*'s method on *Capital*. Arthur argues for a complete homology between the categories in *Capital* and those of *The Science of Logic*. However, as Lange argues, the relationship between *The Science of Logic* and *Capital* is better understood as methodological. *The Science of Logic* influences *Capital* most clearly in its method, which appears in its categorical development and not the presentation of categories. This criticism of Arthur is useful to clarify a correct reading of Marx: *Capital* is not a reconstruction of Hegel's categories. Rather, Marx is concerned with the different levels of abstraction internal to the concept at different levels of the exposition. At the beginning of *Capital*, the concept does not reveal itself but rather exists in the state of its appearance.

---

38   Arthur 2002, p. 85.

To employ Hegel's method as a critique of political economy, Marx fundamentally changes the nature of the categories by attributing them to that of things such as commodities, which is in stark contrast to Hegel's presentation of categories. Hegel's categories do not signify things at all, at least not until 'The Doctrine of Essence' where things are only thematised along with thinking 'existence'. Hegel's things do not exist practically as objects that extend in space or in time. Marx, in contrast, in his method, integrates being and essence in the concept from the very beginning, and this is done through the concretisation of abstractions in things.

When Hegel arrives at his last book, 'The Doctrine of the Concept', the elements within the first two books become understood in relation to the fully developed concept. This is where Marx takes his methodological departure. 'The Doctrine of the Concept' is a point of reference from which he can understand capital holistically. Capital can thus be grasped as a fully developed concept and as self-reproducing (and not as the two-fold articulation of being and essence). 'The Doctrine of the Concept' is the general formula for capital; therefore, it is applied to capital's process of valorisation as well as the general process of capital's reproduction at the level of the whole (and not individual capitals). In the relationship between simple reproduction and expanded reproduction, being and essence are superseded by capital's self-causing unity as subject.[39] It is the unity of production and reproduction that give capital its subjectivity as a self-causing unity with a life process. This is a conceptual moment that cannot be reached without methodologically emphasising 'The Doctrine of the Concept'.

## 5.2  Intersubjective Structures

Following from the establishment of capital as subject – a fully developed, self-reproducing concept – this section addresses its interconnection with the subjectivity of the bearers of the capital relation by examining the concept of 'substance'. 'Substance' functions as the logical mediation between the life of capital and a positive concept of life. It is, therefore, instructive to consider substance to understand the logical development of a theory of reproduction in Marx that encompasses both concepts of life.

---

[39]  Meaney 2002, p. 113. The twofold relations of being and essence are superseded by capital as a self-causing unity or, to be more precise, as the unity of production and reproduction.

In Hegel's *Logic*, 'subject' cannot be defined without addressing its relationship to – and expression *in* – other central categories, including substance, concept, spirit and *idea*. All of these categories are revealed to be interrelated in the second volume of *The Science of Logic*, "The Subjective Logic". Each category is involved in the development of a subsequent category that, for Hegel, is more real than its predecessor yet also contains within it the categories that came before in previous moments. Hence, spirit [*Geist*] is not counterposed to its object (the categories in which it engages) but overreaches them as moments within its own development. *Geist* becomes its categorical derivations and develops into intersubjective structures. Thus, in the well-known Hegelian dictum, the 'I that is we and the we that is I', each category is produced in its intersubjective relation to other categories. In contrast to *Geist*, subject is withdrawn into itself and underlies and is counterposed to the object. However, the subject develops into *Geist* to become the *idea*: the unity of subjectivity and objectivity. Subject is in its first moment individuated as a 'self-referring universality'.[40]

The first moment of the development of the subject, where the subject appears as an individual, or individuated, is a step in the process that is methodologically important for reproduction in Marx's exposition. This moment of individuation is manifest in Marx's concept of capital as subject in the form of appearance of 'money as money,' or as we saw in Chapter 2, 'Money Form', where money's independence as an independent expression of value is secured by credit money. The form of appearance of 'money as money' reposits the categories of the 'living individual' in *The Science of Logic*. As Arthur has observed, 'the triad of capital derives from that of Hegel's "Idea", but somewhat re-positioning its logical categories: The Living Individual, Life, and Absolute Idea'.[41] The living individual is the premise for the reproduction of life, which is a prerequisite for the absolute idea. As an individuated category, the living individual will always remain dialectically other to the absolute, which is an intersubjective processual totality.

The appearance of 'money as money' is prior to that of the life of capital, in much the same way as the category of the individual is prior to life in *The Science of Logic*: the individual is prior to the living individual, which is, analogously, money in motion or in circulation. Because 'money as money' is not in circulation, or motion, 'money as money' constitutes a form of interruption or negativity, and it therefore cannot be reduced to an expression of the cap-

---

40   Hegel 2010, p. 740.
41   Arthur 2022, p. 405.

italist relation of production. As with how life – posited as individual life – is a 'disruption' in the movement to the absolute in *The Science of Logic*, 'money as money' is an interruption from 'money in motion'. This interruption punctuates the self-movement of capital, or capital's life process, and occurs formally. Formally, the interruption of 'money as money' exists as a part of the general *formula* of capital. As such, it is internal to capitalist totality but other to its form.

As discussed in the second chapter, when capital incurs productive barriers to valorisation, 'money as money' stands in for the productive process through its facilitation of circulation.[42] Money as money is an interruption from valorisation and therefore is a form irreducible to capitalist social relations. In contrast, when money is 'money as capital' it is constantly moving towards valorisation. 'Money as capital' becomes 'money as money' when directed towards other purposes than the movement towards valorisation. In this manner, 'money as money' is value individuated: a reposting of the logical category of 'the living individual'.

Money as money circulates as a means of payment or as a fund, and as such is limited to reproductive means and is not a end in itself. Money is indifferent to its uses and thus is individualisable. Here, money is its own aim, permitting it to acquire an individualised life process outside of the aim of capital's life process. As Arthur explains,

> ... the Fund is *indifferent* to its potential uses, and thus it is here isolable as an individuated value. This individuation thus acts as the presupposition for its embarkation on its own 'life', setting M as its aim, not the circulation of commodities.[43]

As in Hegel's insistence in the *Logic* that 'individual life' interrupts the idea as an individuated entity required for its reproduction, money as money is an interruption of capital's self-movement as a necessary component of capital reproduction.

'Life' appears in Section III of the 'Doctrine of the Concept' as the first form of appearance of the idea: the idea in its immediacy. Capital has a life-like character in Marx's methodological reconstruction because it possesses the

---

[42] Money as money, or value individuated, remains a distinct moment in the reproduction of capital, taking the shape of non-valorised value as credit. As non-valorised value, *it cannot be reduced to an expression of the capitalist relation of production*; contradictorily, money as money can only be an independent expression of value in capitalist social relations.

[43] Arthur 2022, p. 181.

capacity to extract abstract labour – socialised homogenous labour – from the sum of living individuals labouring in capitalist society. The life of capital is therefore not the result of the appearance of abstract labour, but abstract labour's effect as an ontological precondition for giving the *idea* the subjective character of life. The *idea* must be rooted in the movement of the concept as organic being. Abstract labour supplies the substance of value, which only appears as value relating to itself in the form of money. Life is therefore an ontological precondition of the reproductive medium of the money form, whose metabolic process (circulation) is the life process, giving subjectivity to capital.

For Marx, as in Hegel, substance, as essence united with being, is intrinsically linked to life or that which is living. However, substance is not the same as life; rather, it is matter or the make-up of something. Yet, in Hegel, as in Marx, the substantial matter is nonetheless derived from that which is living as a condition.[44] The objectification of living labour as abstract labour is the substance of value and, ultimately, that which constitutes the substance of the life-like character of capital as self-moving substance. Here, substance does not correspond directly to life; rather, it is an objectification of life (a fetish character) or the mediation of being and essence (subject and object). The element of life is doubly mediated in the case of value-in-itself (its appearance as money form), dialectically begetting a self-moving substance, which is value in motion, appearing as money in circulation.

The development of the categories of life plays a significant role in the development of the subject in Hegel, specifically in their role 'reproducing' the subject. In Marx's repositioning of these categories in *Capital*, this process becomes more complicated: In Marx, we are not dealing with only one system of reproduction but two. Capital, as the subject of the process, must reproduce itself. But so, too, must human life and nature. In Volume II, Marx refers to the process of capital's valorisation as 'capital's life process'. This development of capital as subject occurs through the movement of the value forms. Marx delineates capital's nature through an abstract formal appropriation of the development of the doctrine of the subject in *The Science of Logic*. In doing so, Marx understands the life process of capital-as-subject through the mobilisation of Hegel's account of life as an essential moment in the development of the subject. As such, Marx's presentation unfolds through a definition of life as a self-reproducing organic unity. Here, money is the medium of circulation and reproductive form of value, providing the formal conditions for capital's

---

44    This is why use values are absolutely necessary; the value form is not immaterial.

reproduction and accumulation. As was established in Chapter 2, 'The Money Form', money can acquire this role only because it is a general equivalent, making it the 'actuality' of value. Money is *value-in-itself* as substance, and therefore also something immanently other to capital.[45] Arthur supports this point when he states that value gains its actuality only when a universal equivalent gains its existence.[46] Money, in its distinction from other commodities, gains its own independent substance. This is why money can be a means of payment, medium of exchange and circulation, and storer of value.

Marx uses the term 'substance' in two different senses: to describe labour as the substance of value and to posit value as substance (money). In both cases, substance is what makes up the life content of capital. Ultimately, however, it is the living component of living labour that produces substance in its reified form as abstract labour. Value as substance is money, the form that circulates and it is the medium of reproduction of capital as a general equivalent. Money is therefore the form of appearance of the self-moving subject of capital. As Arthur notes,

> ... it is important to notice that the presentation of money as 'substance' is a very different use of the term 'substance' from that of Marx when he derives labour as substance *of* value in his first chapter ... [Money as substance is] concerned with value *as* substance (corresponding to Marx's use of the term in a later chapter where he speaks of value in motion as 'a self-moving substance').[47]

With the application of a reconstruction thereof, Arthur's emphasis on Marx's two different uses of substance is an astute clarification. In the first (labour as substance of value), Marx is using substance to describe the constitution of value (its 'stuff' or 'material'). As money, 'substance' obtains an opposite role where the constitution of money is unsubstantial, in the sense that it is merely a 'transubstantiated outer shell',[48] and value itself is the substance.[49] This represents the movement from value-in-itself (value as inner content: the way in which the substance of value is internal to the commodity form) to

---

45  It is a form that is and is not a commodity and is not capital; it is a general equivalent and therefore a form set apart.
46  Arthur 2002, p. 98.
47  Arthur 2002, p. 98.
48  Arthur 2002, p. 99.
49  This is in line with how we have understood money in previous chapters, where it is argued that the money form historically developed out of the commodity gold due to social custom and that this is not at all essential to its role as a form.

value-for-itself (value as outer expression, in actuality in the case of the money form). In the case of the commodity form, or value as inner content, the value 'is purely virtual as its reality is merely the ideality of the unity of the commodities and their abstract identity as exchangeable'.[50] For exchange value to be real measure, value must become an independent expression (in itself). This is necessary to ground the commensuration, for 'in the commodity value is a "quasi-property" while, for money, value is itself a substance with a use-value: therefore, it is the actuality of value'.[51] Furthermore, value in its form of appearance as money is the substance of the self-referring system of the value forms. Living labour, as its source (or ontological precondition), 'lies outside of this self-referring system of value forms'.[52] Hence, that which supplies the substance (the being and essence) of value must also be other to the value form. Correlatively, that which supplies the substance to capital is a special form of value that is also other to capital. Both of these, the reproduction of human life (in metabolic relation to nature) on the one hand, and the reproduction of the life of capital on the other, supply substance because they are self-reproducing.

With both human life and the life of capital, the variable of substance provides the living component through reproduction. Therefore, these elements (human life and the life of capital) ultimately operate – in accordance with arguments expounded throughout this book – as an internal otherness or 'immanent externality' to the value form. Marx makes explicit that living labour needs to be reproduced through the reproduction of human life, which acts as an ontological precondition, or as a medium, for the abstraction of abstract labour. Human life necessarily lies outside of the independent system of the value form. Abstract labour does not lie outside, nor does labour as such; however, the 'living aspect' of living labour does. Labour needs life, but life does not need (wage) labour.

## 5.3 The Category of Reproduction in Hegel's The Science of Logic

This section retrieves Hegel's account of reproduction in *The Science of Logic* with the purpose of employing his account methodologically. This is undertaken to develop a theory of reproduction in *Capital*. In doing so, this section

---

50 Arthur 2002, p. 96.
51 Ibid.
52 Ibid.

reveals that the process of reproduction in *Capital* is a contradictory structure in which life must retain a 'productive externality' within the logic of capital. Life is posited as an immanent, yet negative, external to the concept of capital. Such logical negativity contains a concept of life that has its roots in a positive account of organic being. This section thus reconstructs the logical place of concrete reproduction in relation to the reproduction of capital. Such reconstruction is required for analysis in the final section, 5.4, 'Concrete Reproduction of Human Life and Nature'.

'The Science of Subjective Logic or The Doctrine of the Concept' makes up the third part of the *Logic* and starts with the section 'The Concept in General'. This corresponds to Marx's depiction of the general formula for capital, or capital in its self-movement in expanded reproduction. The first sentence of Hegel's 'The Concept in General' states that the *'nature of the concept'* cannot be given to the reader right away:

> What *the nature of the concept is* cannot be given right away, not any more than can the concept of any other subject matter. It might perhaps seem that, in order to state the concept of a subject matter, the logical element can be presupposed, and that this element would not therefore be preceded by anything else, or be something deduced, just as in geometry logical propositions, when they occur applied to magnitudes and employed in that science, are premised in the form of *axioms, underived and underivable* determinations of cognition.[53]

Therefore, readers must start with a logical element that presupposes the concept: what Hegel refers to as an *absolute foundation*. However, this absolute foundation has to have made itself into an *absolute foundation*. This appears first as an abstraction and therefore is mediated; however, the *absolute foundation* nonetheless retains the character of being immediate, albeit immediate on the basis of a sublimation of the mediation internal to the abstraction. As Hegel explains,

> Now the concept is to be regarded indeed, not just as a subjective presupposition but as *absolute foundation*; but it cannot be the latter except to the extent that it has *made* itself into one. Anything abstractly immediate is indeed a *first*; but, as an abstraction, it is rather something mediated, the foundation of which, if it is to be grasped in its truth, must therefore

---

53   Hegel 2010, p. 508.

first be sought. And this foundation will indeed be something immediate, but an immediate which has made itself such by the sublation of mediation.[54]

What Hegel means by this is that the concept, as an *absolute foundation*, comes after *being* and *essence*, or the immediate and reflection. *Being* and *essence* are moments of the becoming of the concept, yet the concept is also their foundation (and therefore their *truth*). The concept is the result of the unity of being and essence.

In Hegel's account, substance is already 'real essence', or essence united with being, and becomes the concept when actual. Substance is the immediate presupposition of the concept 'substance is *implicitly* what the concept is *explicitly*'.[55] The actuality of the concept (its substance as the unification of being and essence) in 'The Doctrine of Essence' formally mirrors Marx's exposition of capital as coming into being, or how it becomes 'actual', as money form. Analogously, this makes money the form of appearance of the essence of capital: it is *value-in-itself*. As *value-in-itself*, money is the first form of appearance of capital as actual appearing in the last section of 'The Doctrine of Essence' before we arrive at the 'Doctrine of the Concept'.

The immediate genesis of the concept is the dialectical movement of substance through causality and the reciprocal affection in its becoming. Becoming is the reflection of something that passes over into its ground, where the other that this 'something' has passed over to constitute its truth. Therefore, we find that the concept 'is the truth of substance, and since necessity is the determining relational mode of substance, freedom revels itself to be the truth of necessity and the relational mode of the concept'.[56] Substance has therefore posited that which is *in and for itself*. In this way, we find that substance, as the *in and for itself*, is always other to the concept as immanent and external. Hegel relays the concept as the absolute unity of being and reflections in the following passage:

> The necessary forward course of determination characteristic of substance is the *positing* of that which is *in and for itself*. The concept is now this absolute unity of *being* and *reflection* whereby *being-in-and-for-itself* only is by being equally *reflection* or *positedness*, and *positedness* only is by being equally *in-and-for-itself*. – This abstract result is elucidated by

---

54  Hegel 2010, p. 508.
55  Hegel 2010, p. 509.
56  Hegel 2010, p. 509.

the exposition of its concrete genesis which contains the nature of the concept but had to precede its treatment.[57]

In order to produce this abstract result, the concept must have gone through the concrete exposition of its genesis: it is the result of a process. Substance is the actual *in-and-for-itself* (absolute). Substance is in itself because it is the simple identity of possibility, and actuality and absolute because it is the essence containing all actuality and possibility within itself for itself. This is an identity that is absolute power or absolutely self-referring negativity, i.e. subject. Substance is therefore a category that is implicated in the development of the subject and cannot be articulated as independent from the concept of the subject. Rather, it is an internal moment within the development of the subject/concept. Yet it is completely determined by its relationship to 'life' as an internal reproducing other. In this regard, subject has a speculative identity with substance and comes to represent the unity of substance's own self-determination.

In *The Science of Logic*, the transition from substance to subject takes place in the transition from the doctrine of essence to the doctrine of the concept, which occurs in the section on 'Actuality'. 'Actuality' is the purposive activity of form or *formtatigkeit*. Actuality comes into being as an immanent transition from the doctrine of essence to the doctrine of the concept, or transition from substance to subject. This is a self-actualisation that develops within the speculative identity between life and the self-conscious concept, which is always already rooted in life. In Marx, the movement of the value form, or *formtatigkeit*, is the self-actualisation of exchange value (actualisation is internally also a use-value). The actuality of value as the movement of value form [*formtatigkeit*] expresses how value, like the self-conscious concept, is not immaterial. Value's very materiality endows it with the purposiveness of a subject character.

In Hegel's *The Science of Logic*, within the movement from substance to subject – or from necessity to both freedom and self-determination – the process of actualisation is understood in terms of an immanent self-actualisation or purposive activity. Hegel understands the movement of forms as a purposiveness where the genesis of the concept is the actualisation of activity that is purposive: For Hegel, there is a speculative identity between life and the concept. As Karen Ng, a commentator on Hegel's *Logic* who pays close attention to the 'Life' section, lucidly explains,

---

57   Ibid.

> ... the immanent transformation from substance to subject, from necessity to freedom and self-determination, is an immanent transformation that takes place by means of an investigation into the concept of actuality, asking how the process of actualization can be conceived of in terms of self-actualization. In understanding activity of form in terms of purposiveness, Hegel argues that the genesis of the Concept arises from and is an actualization of purposive activity, once again demonstrating the reciprocity and speculative identity between life and the self-conscious Concept.[58]

The movement from substance to subject, immanent self-actualisation or purposive activity (teleology) has a positive role to play in Hegel's logic. By opening 'life' to 'the *idea*', 'actuality' establishes a reciprocity between internal purposiveness and judgment. This is Hegel's epistemological materialism: we cannot merely have a concept of the 'concept of the concept.' The 'concept of the concept' has to pass over into actuality. Hegel therefore establishes a link between life (internal purposiveness) and self-consciousness (judgement).

An account of Marx's presentation is sharpened by retrieving Hegel's exposition of the movement from substance to subject because it demonstrates how abstract labour (rooted in human life), as substance of value, supplies the purposeful character necessary to the transition from substance to subject: in the movement from abstract labour to value. This is only possible due to substance's roots in life. Unsurprisingly, the purposeful character of self-actualisation is also self-reproduction and value only becomes 'actual' in the form of appearance of money, the medium of the reproduction of capital.[59]

Hegel logically understands life as a concrete reality. However, since life is *a priori*, it is not conceptually understood as empirical but as logically holding the place of organic being. For Hegel, life is internal to the subject function of the concept. The category of life appears with the purpose of rooting the concept in organic being. Hegel makes the argument that the other of the concept, internal to the concept, is 'life'. As life, it plays the logical role of a medium of reproduction and at the same time must posit itself in externality. Hegel explains,

---

58  Ng 2020, p. 127.
59  Meaney explains this well in the following quotation: 'Marx describes capital as self-reproducing and multiplying and, as such, perennial. Moreover, he goes on to say that when capital relates itself to itself as self-reproducing, it distinguishes itself within itself from itself as profit, and then supersedes the separation and thereby expands itself as the subject of a self-expanding circle or spiral' (2002, p. 41).

... the unity of the concept posits itself *in its externality* as negative unity, and this is *reproduction*. – The two first moments, sensibility and irritability, are abstract determinations; in reproduction life is *something concrete* and vital; in it alone does it also have feeling and power of resistance. Reproduction is the negativity as simple moment of sensibility, and irritability is only a vital power of resistance, so that the relation to the external is reproduction and identity of the individual with itself.[60]

Life is as much a part of the concept as it is other to the concept, an externality or 'negative unity' logically made so on account of the concrete and vital nature of its reproduction. For Hegel, the *idea* – to be self-moving – requires life (individual life, the life process and genus) as its necessary internal other that drives movement through reproduction. This is why in Marx, when capital is seen as self-moving, it is self-moving because it is able to reproduce itself and therefore has a 'life-like' function – what Marx calls its 'life-process'.

As the second moment in the development of life from individual life to genus, the life process is determined in Hegel first and foremost by reproduction. Reproduction is understood as at once the individual relation to the external and the positing of the concrete totality of the whole. For Hegel, reproduction is 'a moment of singularity' where an individual becomes actual and in becoming actual is relational within a concrete totality:

> With reproduction as a moment of singularity, the living being posits itself as *actual* individuality, a self-referring being-for-itself; but it is at the same time a real *outward reference*, the reflection of *particularity* or irritability *as against an other*, as against the *objective* world. The life-process enclosed within the individual passes over into a reference to the presupposed objectivity as such, by virtue of the fact that, as the individual posits itself as *subjective* totality, the *moment of its determinateness, its reference* to externality, also becomes a *totality*.

While,

> ... each singular moment is essentially the totality of all; their difference constitutes the ideal determination of form which is posited in reproduction as the concrete totality of the whole.[61]

---

60  Hegel 2010, p. 683.
61  Hegel 2010, p. 683.

Hegel's account of reproduction in the life section brings clarity to the consequences of Marx's description of capital as having a 'life process'. Marx reflects that life has a dialectical relationship to the movement of the value form that requires negativity: the positing of an externality that is simultaneously immanent to the totality of the relations internal to the development of the concept (capital). There is a contradictory structure within the social universality of the relationship between capital and life. Life retains a 'productive externality' that exists within the process of valorisation.[62] The process of reproduction requires the positing of a negativity external to the immediacy of the concept. This negativity is a concept of 'life' that has positive roots in organic being. Hence, the category of reproduction in Hegel anticipates Marx's development of capital's 'life process', or of capital as subject.

Marx's inheritance from Hegel, in his deployment of reproduction, makes it clear that the *idea* anticipates Marx's understanding of capital as a life process. This unfolds within the dynamic of reproduction in Volume II of *Capital*. Grasping this relation, reveals the significance of the enigmatic use of 'life' in *Capital* Volume II. 'Life' as an agent of reproduction, confirms the development of the value form as an automatic subject. This is proven to be possible only because reproduction is a contradictory process that contains its own negation. Hegel's deployment of reproduction in his understanding of 'life' makes clear that 'the living being is the externality of itself as against itself'.[63] The living contains an 'immanent reflection' that sublates its own immediacy.

Attention to Marx's emphasis on capital's life process illuminates the nature of the central contradiction sustaining its concept. Instead of isolating labour as the central contradictory other of capital – which is in fact not only a part of the process of valorisation, but also a product thereof, representing only one form of the subjection of human life within a larger social universality where human actors have different personified roles – we can interpret the central contradiction of capital's reproduction from the point of view of an ontological precondition: life. Owing to Marx's use of the life process, appropriating the reproductive function of 'logical life' from *The Science of Logic*, the dynamics of 'life' thus constitute a pivotal philosophical feature of *Capital* Volume II.

In *The Science of Logic*, as organic form, life inherently posits its own externality from the concept as a necessary condition for the reproduction of the 'concept' and its subsequent development as 'actual'. A non-dialectical speculative element of freedom (rooted in organic being) is behind the movement

---

62  Which is a process of reproduction.
63  Hegel 2010, p. 683.

from life in its immediacy to self-conscious life. This results in a process of subjectivation that also appears as a result of the speculative element within Marx's exposition, where the *idea*, or capital, contains within it an element of freedom or 'self-movement'. This element of freedom or 'self-movement' is sustained by its rootedness in the non-dialectical purposiveness of organic life.[64] Reproduction for Hegel requires a 'moment of singularity', or the individual self-positing of life, and

> ... thus the idea is, *first of all life*. It is the concept which, distinct from its objectivity, simple in itself, permeates that objectivity and, as self-directed purpose, has its means within it and posits it as its means, yet is immanent in this means is therein the realized purpose identical with itself. – The idea, on account of its immediacy, has *singularity* for the form of its concrete existence. But the reflection within it of its absolute process is the sublating of this immediate singularity; thereby the concept, which as universality is in this singularity the *inner*, transforms externality into universality, or posits its objectivity as a self-equality.[65]

This statement articulates the external ontological function of life as self-directed purpose, which, when combined with Hegel's avowal that 'the unity of the concept posits itself *in its externality* as negative unity, and this is *reproduction*'[66] confirms three essential points at the centre of our interpretation of reproductive forms' generation of immanently external social practices:

1. Life, as a moment of singularity, or individual self-positing, is the purposeful precondition behind the speculative nature of the *idea*, where self-directed purposiveness 'permeates' objectivity immanently (as in the theory of money as money).
2. This purposiveness or the element of freedom required for the development of the concept (its speculative element) must be rooted in externality or individual self-purpose (the immediacy of the *idea* is a singular form of concrete existence).
3. This externality is transformed by the absolute process or universality of the concept and is posited in self-equality with the concept, rendering its externality immanent (money in motion, or the general equivalent, takes on this role in Marx).

---

64   The conditioning of abstract labour is labour.
65   Hegel 2010, p. 675.
66   Hegel 2010, p. 683.

Attention to the speculative element of the self-movement of capital – which is possible through its rootedness in the non-dialectical purposiveness of the immediate *idea*, or life – enables access to an analysis of the social mediations that uphold the value form as contradictory structures of subjection and subjectivation. These social mediations entail social practices that not only function to reproduce the movement of the value form but also to reproduce individuals' lives in practice.

## 5.4  Concrete Reproduction of Human Life and Nature

Departing from the claim that Marx's theory of concrete reproduction is left partial and requires reconstruction, this section theorises the logical place of the concrete reproduction of human life and nature. This is done by first establishing the contradiction between human life and nature's transhistorical metabolism and capital's abstract form to be an irresolvable antagonism. This is shown to be a result of the ontological differences between Marx's two concepts of life (the difference between abstract reproduction and concrete reproduction is essential to the ontological difference in Marx's two concepts of life). Next, this section contends that what constitutes Marx's positive sense of life requires further determination. The need for further determination is then addressed by employing Marx's anthropology of practice, wherein human essence is interpreted not only as an ensemble of social relations, but also as social relations metabolically related to nature; ecology is thus established as essential to 'life'.

Situated within the legacy of German idealism and the burgeoning literature on the natural sciences, Marx reveals his roots in a combination of his critique of Feuerbach and of physiology with particular focus on anthropology as a means to define concrete life. Thus, it is argued here that Marx's critique of Feuerbach works to establish a framework that makes room for the inclusion of extra-philosophical discourses, drawing on positive sciences to develop an anthropology and relational exposition of matter and the natural world. This section thereby establishes the basis for the conclusion of this chapter, wherein a positive concept of life is used to reconstruct Marx's theory reproduction of capital relations as a whole, where concrete reproduction is found to be both immanent and external to the reproduction of capital as an abstract form.

Upon consideration of the practices that reproduce life, we find influences of non-capitalist variables and empirically given conditions,[67] which include

---

67  Marx 1991, pp. 927–8.

interpersonal forms of domination, natural limits and vast amounts of unvalorised value in circulation (money as money, or credit money). These work to determine human life and are dialectically both potentially transhistorical and determined by capital's abstract forms. Here, the essential contradiction of capital is one between the transhistorical metabolism of human life and nature (nature referring to the sensuous world, or the in-organic body or material of capital's abstract forms) and capital as a process of circulation that has productive and unproductive elements.

So far, we have discussed Marx's method and corresponding insights into how to understand his demonstration of economic reproduction through Hegel's logic of the *idea*. However, this provides insight into the logic of form analysis only. Capital as an abstraction is 'the free subject with its independent right'.[68] Here, Hegelian reconciliation of being and essence in the concept occurs only within the movement of abstractions, and there is no reconciliation between matter and form. Therefore, matter cannot be ontologically understood from the point of view of Hegel's philosophy. This means, philosophically, that Marx advocates a double ontology. There is the ontology of capital and its reified forms, and there is an ontology of human life and nature in their biological and historical modes. This antagonism between value form and life is a practical objective tendency positioned in oppositional unity, and not a logical contradiction. In this way, for Marx, Hegel's logic remains at the level of an ontology of capital, or a logic of abstractions, and does not provide a basis for ontological understandings of human anthropology or matter and the natural world.

Due to natural limits, the metabolic process between humans and nature will always remain antagonistic to capital's social forms. But material will also be historically altered, as in any epoch. This alteration takes on a particular character in capitalist social relations. Therefore, close attention to Marx's form analysis ought not to be used merely to grant ontological primacy to capital's abstractions, but it should be analysed as a materialist orientation for under-

---

68  Arthur 2000, p. 122. A rejection of this claim might contend that capital does not fulfil the reconciliatory Hegelian logic of the *idea*, because there is an antagonism between living labour and capital. However, this assumption fails to see that the concrete aspect of labour remains outside of the formal movement of capital's abstract form. Therefore, within this self-movement of capital there are internal antagonisms between itself and its other that go on to achieve reified reconciliation as capital. This is a central attribute of the category of abstract labour. Reconciliation occurs as a result of the process of capital's formal realisation, and the moment of reconciliation – necessary for a meaningful application of Hegel's philosophical system – takes place in an abstract form only.

standing what the precise limits of capital's abstractions are and how they relate to the complexity of materials that constitute the totality of the world.

Lise Vogel has claimed that, in *Capital*, Marx overlooked and excluded how theorisation of the reproduction of the labourer could be demonstrated and therefore his theory of reproduction was left partially developed. In contrast, I claim, through the interpretation deployed in 'Immanent Externalities,' that Marx's development of capital's reproduction is not partial simply because he missed a theory of how labour is reproduced. While Marx's exposition logically worked out the abstract nature of capital's reproduction (revealing Hegelian logical roots), what was left partial was concrete reproduction more broadly. The partial nature of concrete reproduction is not only due to lack of attention to the reproduction of the labourer, but also the physiological and ecological aspects of concrete reproduction, which would require studies in natural science as well as epistemology to clarify how one can know what concrete life is.

The degree of exclusion of concrete reproduction in *Capital* might be attributed to the fact that the non-capitalist aspects of capital – which practically make up concrete reproduction – are local, contingent and messy historical realities that cannot be reduced to neat concepts. Concrete reproduction will also have immanently capitalist forms as capital's abstractions modify the concrete dimension of the world. While concrete reproduction is partially worked out within Marxist literature and twentieth- and twenty-first-century readings of *Capital*, both within social reproduction theory and studies of Marx's ecology, the logical function of the latter positive concept of life – where biological life functions as a negation of abstract life – is little understood.

This interpretation broadly aligns with a 'unitary theory' of reproduction, where capital's abstractions are understood to play a role in the reproduction and conditioning of non-capitalist social relations. However, this chapter's interpretation also aims to sharpen an understanding of the dialectical relationship between the reproduction of capital and the reproduction of biological life to the extent that I claim there are ontologically contradictory concepts of both life and reproduction in Marx. As ontologically contradictory concepts stemming from concepts of life internal to Marx, reproduction is an 'immanent externality', or a negative dialectic within capital's logic. As logically other to capital, and a negation thereof, the positive concept of reproduction (internal to Marx's positive concept of life) becomes a potential site of resistance.

From here, it is possible to infer an internal displacement within Marx's critique that follows his development of two different concepts of 'life'. One concept is Hegelian, which works within *Capital* to explain the life of capital as an abstract form of social domination. The other concept derives from Marx's own disjunctive project, which included two sides that were never distinct: a critique of Feuerbachian humanism and an intensive study of the natural sciences. This unfinished, disjunctive project within *Capital* expresses that the concept of life is derived from the concrete, and, although it operates as a negation to the life of capital, it is not merely thought in terms of a negative ontology.

Marx's positive sense of life is a negation to the life of capital and therefore exists in permanent tension with the life process of capital, possessing a distinct ontology. Thus, this positive sense of life provides a basis for possible resistance against capitalist social relations. However, to adequately address the way this positive sense of life acquires its determinacy is not the purpose of this book. Instead, this book aims to construct a logical account of the function of the positive concept of life in relation to capital's abstract life process. Nonetheless, it is important to approximate the determination Marx gives to life in order to develop a plausible logical category.

Marx refers to a positive account of life when referring to the life of the worker in the following passage, which that exemplifies his use of the term: 'the consumption of labour-power by capital is so rapid that the worker has already more or less completely lived himself out when he is only half-way through his life'.[69] Marx also repeatedly refers to a concept of human life in a physiological sense throughout *Capital* in his earnest descriptions of the exhaustion and exertion of human muscles, bones, brains and nerves:

> The capital given in exchange for labour-power is converted into means of subsistence which have to be consumed to reproduce the muscles, nerves, bones, and brains of existing workers, and to bring new workers into existence.[70]

These descriptions retain a strong empirical commitment to human experience, and they make evident Marx's preoccupation with the way in which capital curtails the potentialities of humans as free purposive individuals, is also clear in his claim that 'capitalist production has seized the power of the people at the very root of life'.[71]

So far, I have elaborated on Marx's interpretation of the life of capital as a metabolism of abstract forms or the 'movement of value form'.[72] The move-

---

69   Marx 1990, p. 795.
70   Marx 1990, p. 717.
71   Marx 1990, p. 380.
72   In the original German, 'movement of value form' is referred to as *formwechsel*. The Ger-

ment of value is theorised through an anti-empirical critique, where abstractions are thought to mediate appearances made up of matter. These appearances exist in contrast to the essence of what they are: essence is here determined by abstractions' mediation. In this dynamic, capital's economic form determinations modify material; however, this material nonetheless retains its own qualities and limits. There are also physiological and ecological factors that exist in tension with and place limits on the movement of capital's abstract forms.

The logical relation of the physiological and ecological factors that exist in tension with and place limits on the movement of capital's abstract forms were discussed in Chapter Four, 'Time and Schemas of Reproduction'. The chapter considered how, in Marx's schemas of reproduction, material or matter determined the timing of circulation by placing natural limits on rates of circulation and accumulation. Materials, for instance, degrade and need to be reproduced at different rates depending on what materials they are. Matter determines capital's ability to renew its conditions of production through reproduction. Put another way, the period of reproduction, or time required for the renewal of capital, is based on material conditions determined by natural properties of the material. This includes both the renewal of labour power and the renewal of fixed and circulating capital. The material bearer of capital is therefore active and not indifferent to capital's abstractions. For example, it is the constitution of the physical entity itself that determines the difference between fixed and circulating capital. From this perspective, it is obvious that, due to their material property, use values impose differences in the period or time in which they require reproduction, and in doing so condition capital's renewal.

It is thus evident that material bearers – whether human, wooden, technological or otherwise – condition the accumulation of capital. This is because concrete life is articulated as force opposed[73] to the abstract movement of capital's value forms. However, it is far from clear how this sense of life acquires determinacy.[74] Although Marx turns to the natural sciences to interpret the

---

man term is contrasted with *stoffwechsel*, which is translated as 'metabolism' or the 'movement of matter'.

73    This force is made up of both nature and humans who purposively mediate nature through labour.

74    Commentators are conflicted on this point, with two predominant standpoints. On the one hand, it has been claimed that Marx's use of nature in *Capital* contains a remanence of Feuerbachian essence combined with natural-science materialism – a philosophical rationality that is the basis of Alfred Schmidt's influential study, *The Concept of Nature in Marx*. Meanwhile, others claim that Marx turns away from philosophy to engage with natural sciences from an anti-philosophical empirical perspective by the time he writes

concrete aspects of his economic analysis – to determine natural limits within capitalist societies – this aspect of his research is only partially present across the three volumes of *Capital*. As such, what is present is imbued with a hybridity of determinations, where the construction of a positive sense of life draws upon a combination of natural science and the critical application of Feuerbachian humanism.

Marx's intention was not to produce a negative ontology, where nature and matter are only negations of capital. This is why he studied natural history in detail.[75] Marx's investment in natural historical research compels philosophical exegeses to question the extent to which he relied on unmediated empiricism when utilising scientific thought to explain physiology and ecology. Likewise, it is imperative to consider the extent to which Marx remained epistemologically critical when thinking about the underpinnings of matter's appearance.

In *Capital*, Marx's anthropology differs from and is increasingly critical of his earlier humanistic view of human life (tied to his theories of alienation), which evolved from Feuerbachian ideas of life and human nature, developed in *The Essence of Christianity*.[76] Feuerbach, relying on the premise that humanism equates to naturalism, theorised human species-being as an essence composed of freedom expressive of thinking, loving, and willing. This essence was

---

*Capital* – this is the claim made by Saito in his book, *Karl Marx's Ecosocialism*. The interpretation herein finds neither perspective wholly convincing. See Saito 2017; Schmidt 2014.

75  Marx states in the economic manuscript of 1864–5: 'actual natural causes for the exhaustion of the land, which incidentally were unknown to any of the economists who wrote about differential rent, on account of the backward state of agricultural chemistry in their time' (Marx 2015, p. 768). This passage reflects the level of importance Marx gives to the natural sciences. As deeply evidenced by Saito, Marx read the natural sciences rigorously and was particularly influenced by Justus con Liebig's book *Familiar Letter on Chemistry* and Carl Frass's book Die Natur in der Wirthschaft: Erschöpfung und Ersatz. Both respectively developed theories of metabolism that Marx reinterpreted as a social theory, which is deployed throughout his work, for example in the *Grundrisse* on p. 271 and in *Capital* Volume I on p. 289 (Marx 1973; 1990). Further, in 1865, Marx studied natural sciences in detail when developing his theory of ground rent. Marx wrote the following to Engels on the subject: 'I had to plough through the new agricultural chemistry in Germany, in particular Liebig and Schönbein, which is more important for this matter than all the economists put together, as well as the enormous amount of material that the French have produced since I last dealt with this point. I concluded my theoretical investigation of ground rent two years ago. And a great deal had been achieved, especially in the period since then, fully confirming my theory' (Saito 2017, p. 153).

76  Marx's development of his understanding of human life as taken up in *Capital* begins in *The German Ideology* (the "Theses on Feuerbach" was intended to provide the basic outline of the latter text that grew into a larger project).

thought to be both natural and transhistorical. The particularity of this natural human essence is that it is based on consciousness as both a mode of being and an object of thought. In this way, humanity for Feuerbach has a two-fold life where there is both an inner and an outer life, both an 'I' and a 'Thou'. Because, here, one's individuality is an object of thought, individuals are said to relate to themselves along the same structure in which they relate to others collectively. Feuerbach develops this understanding of human essence in order to show that religion is the object of this characteristic at the level of the infinite. He begins with the claim:

> ... but what is the being of man of which he is conscious, or what is that which constitutes in him his species, his humanity proper? Reason, Will, and Heart. To a complete man belongs the power of thought, the power of will, and the power of heart. The power of thought is the light of knowledge, the power of will is the energy of character, the power of heart is love. Reason, love, and power of will are perfections of man; they are his highest powers, his absolute essence in so far as he is man, the purpose of his existence. Man exists in order to think, love, and will. What is the end of reason? Reason. Of love? Love. Of will? The freedom to will. We pursue knowledge in order to know; love in order to love; will in order to will, that is, in order to be free. Truly to be is to be able to think, love, and will.[77]

For Feuerbach, thinking, loving and willing are infinite in scope at the level of humanity or the collective human being: this is what constitutes humanity's freedom. These 'powers' are understood as nothing without the objects that express their being. Feuerbach writes,

> ... man becomes conscious of himself through the object that reflects his being; man's self-consciousness is his consciousness of the object [the inner I is a product of its relationship to the outer thou]. One knows the man by the object that reflects his being; the object lets his being appear to you; the object is his manifest being, his true, objective ego. This is true not only of intellectual but also of sensuous objects.[78]

In this regard, the object to which a subject relates themselves becomes nothing other than the subject's objective being. From this premise, Feuerbach develops

---

[77]  Feuerbach 2012, p. 99.
[78]  Feuerbach 2012, p 101.

his critique of religion. Feuerbach thus derives that the idea of God is nothing other than the synthesis of human perfections objectified in a religious social structure through universalised personification.[79] In this understanding, God is thus an abstraction that is not internal to each individual, for 'in its reality it is the ensemble of social relations'.[80] The 'truth' of religion is, therefore, nothing other than the objectification of collective subjectivity, rendering its transcendence illusory. This illusion constitutes a fundamental alienation between sensuous perception and the objective social world, which, for Feuerbach, is false.

Despite attempting to instate a materialist programme with atheist principles attentive to the collective nature of the individual, Feuerbach's understanding of sensuous perception does not overcome romantic subjectivism, i.e. humanism. For Marx, this constitutes a problematic limitation, because Feuerbach's analysis remains merely epistemic in its scope; it therefore explicitly remains abstract as 'philosophical utopianism', lacking a materialist social ontology.

For Marx the illusion of something like religion will not disappear simply because its falsehood has been recognised: a critique of practice that addresses material objectivity is needed. Marx insists that pure sensuous perception guaranteeing our ability to access essence independent from objective social relations does not exist. Instead, sensuous perception is conditioned by a practical engagement with the world as a process of co-determination. In this manner, Feuerbach's sensuous perception remains an abstraction because, while the social conditions are determined by sensuous perception, he fails see that perception is also determined by social conditions. This commitment to essence inevitably curtails praxis by naming the so-called reality without a method for changing it. Praxis is therefore hindered by Feuerbach's satisfaction with discovering 'essence', where he gives expression to alienated reality whilst leaving it unchanged.[81]

Marx, by contrast, locates Feuerbach's essence itself as an expression of alienated reality. Marx contends that a critique of objective social relations is required to comprehend how contradictions within society might be overcome and not merely pointed out. This does not mean that Marx abandoned any inquiry into human sensuous perception. Marx was committed to the development of an ontology of human life as sensuous; however, he was committed to doing so without a reliance upon pure epistemology as its basis. Instead, Marx turned to practice.

---

79   Balibar 2014, p. 27.
80   Marx 1994, p. 100.
81   Saito has similarly made this point in *Karl Marx's Ecosocialism* (2017, p. 56).

It is instructive to recall the sixth thesis on Feuerbach that sets the ground for Marx's critical engagement with essence:

VI  Feuerbach resolves the religious essence into the human essence. But the human essence is no abstraction inherent in each single individual.
In its reality it is the ensemble of the social relations.
Feuerbach, who does not enter upon a criticism of this real essence, is consequently compelled:
1. To abstract from the historical process and to fix the religious sentiment as something by itself and to presuppose an abstract – isolated – human individual.
2. Essence, therefore, can be comprehended only as "genus", as an internal, dumb generality which naturally unites the many individuals.[82]

Here, Marx, in moving closer to his programme within *Capital*, rejects that there can be essence independent from the social. In doing so, he indicates that human essence is an ensemble of social relations. This provides the starting point for Marx's anthropology in *Capital*; however, it does not yet fully evince the more developed ideas in *Capital* that are rooted in his study of physiology. Initiated by his commitment to the concrete, studies in physiology led Marx to theorise human essence not only as an ensemble of social relations, but also as an ensemble of practices that have metabolic relations with ecology.[83]

Marx defines human essence in the following passage from Chapter Seven of *Capital* Volume I, 'The Labour Process and the Valorization Process':

The labor process ... is purposeful activity aimed at the production of use values. It is an appropriation of what exists in nature for the requirements of man. It is the universal condition for the metabolic interaction [*stoffwechsel*] between man and nature, the everlasting nature-imposed condition of human existence, and it is therefore independent of every form of that existence, or rather it is common to all forms of society in which human beings live.[84]

---

82  Engels and Marx 1975, p. 100.
83  Marx's starting point for this reconceptualisation is a double inquiry into political economy and natural history. This became an endeavour that lasted the rest of his lifetime and left behind the unfinished manuscripts of Volumes II and III of *Capital*. See Saito 2017, p. 61.
84  Marx 1990, p. 290.

As this passage reveals, opening with a description of labour as 'purposeful activity', Marx's rejection of internal purposiveness does not mean that humanity is devoid of purposiveness, consciousness or even freedom. Rather, life is understood as a practical result of a metabolic relation, based on a combination of physiological interactions, practices and, in the case of humans, conscious interactions that constitute the whole of social and material relations. Labour, which mediates humans and nature, is a conscious and purposive interaction with the external sensuous world that generates life practically. Therefore, Marx's critique of Feuerbach – his rejection that essence is anything other than social relations – enabled him to bring to the fore ecology and the natural sciences as a central component of all life functions.

The consequences of Marx's critique of Feuerbach enrich the three Volumes of *Capital*, where living variables play a special role because that which is living reproduces itself and, in doing so, always relies on an interaction with the external sensuous world. Therefore, on the one hand, life can never be fixed; on the other, life develops in relation to the world. The living being and the sensuous world are, as such, interdependent – due to natural limits – and co-determining – on the basis of physiological interactions of production, consumption and excretion.[85] Marx's sense of the living has a teleology, or purposefulness, towards its own reproduction only within an ensemble of relations that are social and natural or engaged in metabolic processes. It is in this regard that Marx's use of the term 'life' is explicitly materialist. Yet, at the same time, his materialism is derived from the legacy of German idealism, which coupled life (derived from nature) with a concept of purposiveness.[86] When this purposiveness meets human consciousness, it is inextricably linked to the question of human freedom. Marx contributes to the legacy of German idealism by conceptualising human essence's underpinning in practical activity. This development of human essence through practice is correspondingly intrinsic to the development or transformation of the world. This, for Marx, 'is a real problem of life'[87] where what we do determines who and what we are. This is why Marx can say that 'capitalist production has seized the power of the people at the very root of life'[88] – because capitalist production undermines the capacity to act (and therefore be) independent of the abstract form determinations that capital allots to individuals.

---

85   Living variables are mediums of reproduction.
86   This refers to the tradition that departs from Kant's articulation of purposiveness.
87   Engels and Marx 1976, p. 302.
88   Marx 1990, p. 380.

## 5.5 Marx's Two Concepts of Life

With the proposition that human essence is derived from practice, Marx applies German idealist discourses of the purposive nature of life – together with Feuerbach, whose legacy broadly draws on this context – in a manner that changes their meaning, negating a utopic sense of essence as derived from nature. At the same time, Marx applies Hegel's idealist use of life as a materially grounded moment within the development of the concept of capital. Marx sees Hegel's account as applicable only to the reified social form of capital, and therefore the idealist concept of life is extended to pertain to the mode and function of the 'life of capital'. Thus, when referring to Marx's understanding of 'life' in *Capital*, there are two meanings – one concrete, the other abstract.

For Marx, life is historical because it is mediated with the objective world through practical activity or 'labour'. Labour for Marx is practice that mediates the external world with the interior world. Labour's mediation of the external and the internal sustains, as both sensuous and thinking, not only the metabolic physiological makeup of the organic being through the production and consumption of means of subsistence, but also 'human enjoyment' and intellectual and aesthetic purpose. In this regard, in the context of different modes of production, humans become different kinds of beings. Nevertheless, life will remain a precondition behind historical difference. Physiological matter will at once be determined by practice, imposed by historical context, and at the same time will always metabolically reproduce itself, regardless of the mode of production that objectively merges with one's organic sensuous matter to produce the specificity of life in a particular historical context.

Marx locates the essence that constitutes human life in labour and, in doing so, radically displaces the source of life from the realm of the transcendent to that of the immanent. In locating essence in praxis, essence disappears, as transhistorical and 'natural', without the relinquishment of other idealist concerns related to human thought, freedom and history. For Marx, praxis as sensuous activity is essence, and it is this that endows humanity with an intersubjective ego (an actor with the capacity to act, as in capital's subjectivity as 'automatic subject'). It is from this perspective that Marx's 'praxis' replaced Feuerbach's 'essence'. In doing so, Marx withdrew human life from German idealism's conception of a transhistorical standpoint and reinterpreted it as an ensemble of practical social relations that result from the practical mediation between human life and life's inorganic form[89] in a metabolic process.

---

89  This refers to nature/material.

In no way, however, does Marx dispense with the conscious subject of idealist thought. Rather, the subject becomes the result of practice. The subject – the position of the individual endowed with human essence – is no longer epistemological, and therefore it is no longer purely transhistorical. The subject cannot have a source that is completely external (essence in Feuerbach's sense); instead, its source must be immanent to it.

There is an overriding formalism internal to Hegel's idealism that is in tension with the meaning physiology gives to Marx's account of concrete life and its reproduction. The lack of reconciliation between capital's abstract self-movement and concrete metabolic processes resists the subsumption of concept and matter necessary for Hegel's project. Nonetheless, Marx does not abandon an idealist philosophy of life entirely, but retains its commitment to thinking of natural sensuous being as productive of subjective knowledge (occurring form the standpoint of human freedom). For these reasons, Marx's 'materialism of practice' is an unparalleled contribution to the idealist tradition, where life, as immanent, is intrinsically linked to the purposiveness of every social system.

Marx makes these conceptual gains by thinking a permanent dialectical tension between the infinite (the life of capital's self-movement) and the finite (concrete life), which is manifest in his rejection of totalising formal dominance over life. This generates a methodological aporia concerning the two sides: how are we to think both the life of capital and of human life and nature? This unresolved tension suggests that Marx's materialism of practice requires an epistemology of concrete life, for which Marx's own method cannot account. Hence Marx's materialism of practice – far from neatly resolving epistemological difficulties – opens up the terrain for much more thorny problems related to political economy's mediation with the positive sciences (which were shown to be a methodical necessity in their ontological otherness).

Locating the incompatibility of political economy's mediation with the positive sciences is nonetheless productive. It enables analysis to see that concrete life will always retain an element that reproduces itself for its own sake and, in doing so, will engender natural limits that in turn impose limits on capital's abstract forms. At the same time, concrete life will be determined and curtailed by capital's abstractions and the reproduction thereof. This expresses a permanent tension in Marx owing to an ontological distinction that cannot be overcome in any simple way. The application of this reading of reproduction in Marx is politically significant because it permits analysis to question the extent to which concrete life is modified by capital's abstract forms and the extent to which concrete life retains its independence. A study from the standpoint of this logical exposition of reproduction can therefore bring theory and praxis closer to reckoning with the ways concrete life can overcome its role in creating the conditions necessary for the reproduction of capital's abstract forms.

## Conclusion

To conclude, as in the proposition expanded in Chapter 1, 'Fictitious Capital and the Return of Personal Forms of Domination', an interpretation of *Capital* focusing on reproduction can illuminate experiences of contemporary capitalist exploitation elided by production-centric theorisations that dominate Marxist analysis. By considering the development of capital within the post-Bretton Woods world, a theory of how the circulation of money facilitates the reproduction of capital has been developed. I have argued that the reproduction of capital is a system of monetary reproduction, because money opens and closes its cycle. Human individuals are subjected to the circulation of money not only through wages or pay, but also, increasingly, through debt or a means of production funded by debt. Furthermore, it has been found that when money circulates as credit money, it acquires a distinct temporality, placing future constraints on subjects. Therefore, due to the increased circulation of fictitious capital, social domination is not always best understood through the wage relation, nor purely through an analysis of the abstract domination of the value forms.[90] This is especially the case because the lack of limitation associated with the accumulation of fictitious capital undermines and destabilises (through privatisation and other tendencies) the practices of social reproduction upon which it relies. Financialisation has decimated social-reproductive functions. Therefore, social domination is best grasped with consideration of how practices other to capital are posited as a means to reproduce the capital relation as a social totality. Here, the money form subjects human individuals to capital accumulation through finance as well as access to wages (and the lack thereof).

---

90   This contrasts with value-form readings of *Capital* from Moishe Postone and many other contributions where the link to Hegel's logic is developed. Instead, my argument aligns with Lise Vogel when she states in the appendix to *Marxism and the Oppression of Women* that 'most households contribute increasing amounts of time to wage-labour, generally reducing the amount and quality of domestic labour their members perform. Other households are caught in persistent joblessness, intensifying marginality, and an impoverished level and kind of domestic labour. Here, it could be argued, the reproduction of a sector of labour-power is in question' (2014, p. 198).

The theoretical framework expounded, which examines the tension between the reproduction of social life and the reproduction of the life of capital, can help us to understand questions that remain unanswerable to both labour-centric and purely abstract-formal accounts produced by value theoretical readings of *Capital*.[91] A labour-centric approach is generally taken by social reproduction theory, which, while not claiming reproductive labour to be value producing, nonetheless focuses on social reproduction as the reproduction of labour power. The value formal approach, by contrast, often lacks an adequate account of how human life and the material world are reproduced in relation to the value form. By mediating social reproductive orientations with value formal analysis, one might begin to answer pertinent questions about capitalist development. Why, in societies with intensified levels of financialisation, do crises in social reproduction unfold? Why are non-capitalist forms of social exclusion sustained by capitalist societies? What are the natural limitations, social and material, to capital's accumulation? While these questions generate different political and historical conclusions, they each require consideration of how reproduction engenders a form of negativity rooted in a heterogeneity that is more decisively incompatible with the capitalist system than the negativity of production-based analyses of class struggle.

In the culmination of my argument, as I turned to Marx's two concepts of 'life', my interpretation opened difficult methodological problems internal to Marx's critique of political economy. There, I addressed how the positive sciences are structurally important as a determinate, methodological other within Marx's critique. Yet, epistemologically, an aporetic issue was unveiled: two incompatible methods have conjoined. The critique of political economy cannot methodologically think the empirical positive sciences, while the positive sciences are necessary to give relevance to the critique of political economy. By contrast, the positive sciences cannot convincingly think the ontology of capital, yet capital impinges on their subject matter. Here, the status of non-capitalist elements acquires philosophical difficulty. How precisely do they uphold or challenge the reproduction of capital's social form, and how would one tell? Answering such questions might require an exit from the methodological particularities of the critique of political economy; subsequently, there might not be an easy route back. Equally, a deconstruction of *Capital* – departing from the undecidable tension between Marx's two concepts of life – might conclude that the Marxist enterprise inadequately accounts for the very thing that makes its systematicity possible.

---

91  The problems and limitations of both labour-centric and formal-abstract accounts were discussed schematically in Chapter 2, 'The Money Form'.

On the other hand, my analysis opens the possibility of resolving these issues by delineating the logical place of the elements that exceed the epistemological framework of the critique of political economy. This, in turn, establishes a framework to begin synthesising logical concepts with empirical detail. In this regard, one might begin to address the practicalities of different 'modes of subjection' or 'socialisation' and practices of reproduction. These are all 'immanent externalities', as they accord to the particular logical tension between the capitalist and non-capitalist elements underpinning capital's reproduction process. In this book, I developed the means to address the epistemological and methodological issues detailed by expounding the logical incompatibility between two methods of the critique of political economy and of positive science.

Although non-capitalist elements place negative limits on capital's social form, they nevertheless possess a conceptual positivity in their own right. Thus, the contradiction between capitalist and non-capitalist elements exceeds conceptual incompatibility and is found to engender practical incompatibility. The life process of capital on the one hand and of human life and nature on the other begets a contradiction between the possibilities of their respective reproduction. This can be clearly seen in the case of the climate crisis, where the reproduction of the capitalist social form has curtailed the sustainable reproduction of a global ecology.

Equally, human life and nature engender natural limits to capital's abstract forms in a contradiction that has led me to consider how concrete life retains an independence from capital. Can the critique of political economy think the specific positivity of life or not? Might such issues entail recourse to a more thoroughgoing philosophy of science? These difficult methodological and epistemological problems reflect how little understood the reproduction of capital is. This is a problem that I have attempted to rectify. To do so, I have ventured to reconstruct the three volumes of *Capital* from the perspective of reproduction. Here, I have argued that finance capital, social reproduction and ecological reproduction work together as logically interconnected, with each medium placing increased strain on the other.

Understanding capitalist reproduction monetarily, I have shown that a critique of political economy undertaken from the perspective of the production-wage-labour relation will only ever provide a reductive reading of the logic of capital's social relations. Instead, I've argued that capital's central contradiction resides in the tension between capitalist and non-capitalist forms and processes. Correlatively, the concept of 'immanent externalities' was mobilised to articulate the logic thereof. I've argued that accounting for the logical coexistence of impersonal relations and non-capitalist interpersonal relations is formally necessary for reproduction to occur. Retrospectively, from the point of view of the final chapter, the dynamic between impersonal relations and non-capitalist interpersonal relations can be seen as symptomatic of the epistemological scope of the critique of political economy: in its inability to address interpersonal forms of domination adequately, the critique of political economy cannot empirically account for the specific ways in which capitalist reproduction is in tension with the reproduction of human life and nature.

To deepen our understanding of capitalist reproduction, this book provides a framework to address issues at the convergence of social reproduction, ecological degradation and finance capital. But in doing so, I have shown there is a methodological contradiction internal to Marx's critique, the stakes of which are implied by its two concepts of life. It follows that resolving the methodological and practical issues generated – to think of human life and natural life in relation to capital – may well require empirical and philosophical analysis to look beyond the critique of political economy. In practical terms, this would look like the forging of an epistemology more adequate to the concrete.

# Bibliography

Albritton, Robert 1999, *Dialectics and Deconstruction in Political Economy*, London: Macmillan.

Albritton, Robert and John Simoulidis (eds.) 2003, *New Dialectics and Political Economy*, London: Palgrave.

Althusser, Louis 1969, *For Marx*, translated by Ben Brewster, London: New Left Books.

Althusser, Louis et al. 2016 [1965], *Reading Capital: The Complete Edition*, translated by Ben Brewster and David Fernbach, London: Verso.

Arestis, Philip and Malcolm Sawyer (eds.) 2007, *A Handbook of Alternative Monetary Economics*, Cheltenham: Edward Elgar Publishing.

Arrighi, Giovanni 1994, *The Long Twentieth Century*, London: Verso.

Arruzza, Cinzia 2014, 'Remarks on Gender', *Viewpoint Magazine*, available at: https://www.viewpointmag.com/2014/09/02/remarks-on-gender/.

Arruzza, Cinzia 2015a, Gender as Social Temporality: Butler (and Marx), *Historical Materialism*, 23, 1: 28–52.

Arruzza, Cinzia 2015b, 'Logic or History? The Political Stakes of Marxist-Feminist Theory,' *Viewpoint Magazine*, available at: https://www.viewpoint-mag.com/2015/06/23/logic-or-history-the-political-stakes-of-marxist-feminist-theory/.

Arthur, Christopher, J. 1986, *Dialectics of Labour: Marx and His Relation to Hegel*, Oxford: Blackwell.

Arthur, Christopher J. and Geert Reuten 1998, *The Circulation of Capital: Essays on Volume Two of Marx's* Capital, New York: St. Martin's Press Inc.

Arthur, Christopher J. 2000, 'From the Critique of Hegel to the Critique of Capital', in *The Hegel-Marx Connection*, edited by Tony Burns and Ian Fraser, London: Macmillan.

Arthur, Christopher J. 2002, *The New Dialectic and Marx's Capital*, Leiden: Brill.

Arthur, Christopher J. 2022, *The Spectre of Capital*, Leiden: Brill.

Baasch, Kyle 2020, 'The theatre of economic categories: Rediscovering *Capital* in the late 1960s', *Radical Philosophy*, 2.08: 18–32.

Backhaus, Hans-Georg 1980, 'On the Dialectics of the Value-Form', *Thesis Eleven*, 1, 1: 99–120.

Baldi, Guido 1972, 'Theses on Mass Worker and Social Capital,' *Radical America*, 6, 3: 3–21.

Balibar, Étienne and Immanuel Wallerstein 1988, *Race, Nation, Class: Ambiguous Identities*, London: Verso.

Balibar, Étienne 1994, *Masses, Classes, Ideas: Studies on Politics and Philosophy Before and After Marx*, translated by James Swenson, New York: Routledge.

Balibar, Étienne 2014 [1994], *The Philosophy of Marx*, translated by Gregory Elliott and Chris Turner, London: Verso.

Balibar, Étienne 2017, *Citizen Subject: Foundations for Philosophical Anthropology*, translated by Steven Miller, New York: Fordham University Press.

Banaji, Jairus 2013, 'Seasons of Self-Delusion: Opium, Capitalism and the Financial Markets', *Historical Materialism*, 21, 2: 3–19.

Bellofiore, Riccardo and Nicola Taylor ed. 2004, *The Constitution of Capital Essays on Volume 1 of Marx's Capital*, London: Palgrave Macmillan.

Bellofiore, Riccardo 2005, 'The Monetary Aspects of the Capitalist Process in the Marxian System: An Investigation from the Point of View of the Theory of the Monetary Circuit', in *Marx's Theory of Money*, edited by Fred Moseley, London: Palgrave Macmillan.

Bellofiore, Riccardo and Tommaso Redolfi Riva 2015, 'The *Neue Marx-Lektüre*: Putting the critique of political economy back into the critique of society', *Radical Philosophy*, 189: 24–36.

Bellofiore, Riccardo 2016, 'Marx after Hegel: Capital as Totality and the Centrality of Production', *Crisis & Critique*, 3, 3: 31–64.

Bellofiore, Riccardo 2016, 'Suzanne de Brunhoff,' *The Royal Economic Society*, available at: https://www.res.org.uk/resources-page/january-2016-newsletter-suzanne-de-brunhoff.html.

Bellofiore, Riccardo and Carla Fabiani 2019, *Marx Inattuale (The Untimely Marx)*, Consecutio Rerum 5, Rome: Edizioni Efesto.

Berti, Lapo, Paolo Davoli and Letizia Rustichelli 2016, 'Marx, Money and Capital: Interview with Lapo Berti, economist, writer for the magazine "Primo Maggio"', Reggio Emilia: Rizofera.

Berti, Lapo 1974, 'Denaro come Capitale,' *Primo Maggio*, 3:4.

Best, Beverley 2015, 'Distilling a Value Theory of Ideology from Volume Three of *Capital*', *Historical Materialism* 23, 3: 101–41.

Bhandar, Brenna and Alberto Toscano 2015, 'Race, real estate and real abstraction', *Radical Philosophy*, 194: 8–17.

Bhattacharya, Tithi (ed.) 2017, *Social Reproduction Theory: Remapping Class, Recentering Oppression*, London: Pluto Press.

Bologna, Sergio 1973, 'Moneta e crisi: Marx corrispondente della New York Daily Tribune, 1856–57' *Primo Maggio*, 1: 1–15.

Bologna, Sergio 2014, 'Workerism Beyond Fordism: On the Lineage of Italian Workerism', *Viewpoint Magazine*, available at: https://www.viewpointmag.com/2014/12/15/workerism-beyond-fordism-on-the-lineage-of-italian-workerism/.

Bonefeld, Werner, Richard Gunn and Kosmas Psychopedis (eds.) 1992, *Open Marxism 1: Dialectics and History*, London: Pluto Press.

Bonefeld, Werner, Richard Gunn and Kosmas Psychopedis ed. 1992, *Open Marxism 2: Theory and Practice*, London: Pluto Press.

Bonefeld, Werner, Richard Gunn, John Holloway and Kosmas Psychopedis ed. 1995, *Open Marxism 3: Emancipating Marx*, London: Pluto Press.

Bottomore, Tom (ed.)1991 [1985], *A Dictionary of Marxist Thought*, 2nd ed., Oxford/Cambridge, Mass.: Blackwell Publishers.

Brenner, Robert 2002, *The Boom and the Bubble: The US in the World Economy*, London: Verso.

Burkett, Paul 2014, *Marx and Nature: A Red and Green Perspective*, Chicago: Haymarket Books.

Burns, Tony and Ian Fraser (eds.) 2000, *The Hegel-Marx Connection*, London: Macmillan.

Callinicos, Alex 2021, 'Hidden Abode: The Marxist Critique of Political Economy', in *Routledge Handbook of Marxism and Post-Marxism*, edited by Alex Callinicos, Stathis Kouvelakis and Lucia Pradella, New York: Routledge.

Canguilhem, Georges 1966, 'Le concept et la vie', in *Revue Philosophique de Louvain*, 82: 193–223.

Cassin, Barbara, Emily Apter, Jacques Lezra and Michael Wood (eds.) 2014, *Dictionary of Untranslatables: A Philosophical Lexicon*, Princeton: Princeton University Press.

Caygill, Howard 1995, *A Kant Dictionary*, Oxford: Blackwell Publishers.

Chakrabarty, Dipesh 2007, *Provincializing Europe: Postcolonial Thought and Historical Difference*, Princeton: Princeton University Press.

Chesnais, François 2016, *Finance Capital Today: Corporations and Banks in the Lasting Global Slump*, Leiden: Brill.

Chibber, Vivek 2013, *Postcolonial Theory and the Specter of Capital*, London: Verso.

Cleaver, Harry 2019, *33 Lessons on Capital: Reading Marx Politically*, London: Pluto Press.

Collins, Patricia Hill 2008, *Black Feminist Thought: Knowledge, Consciousness, and the Politics of Empowerment*, New York: Routledge.

Crenshaw, Kimberle 1989, 'Demarginalizing the Intersection of Race and Sex: A Black Feminist Critique of Anti-discrimination Doctrine, Feminist Theory and Antiracist Politics', *University of Chicago Legal Forum*, 1989, 1: 139–67.

Davis, Mike 2018, *Old Gods, New Enigmas: Marx's Lost Theory*, London: Verso.

de Brunhoff, Suzanne 2015 [1976], *Marx on Money*, London: Verso.

de Brunhoff, Suzanne 1978, *The State, Capital and Economic Policy*, London: Pluto Press.

de Brunhoff, Suzanne 1979, *Les Rapports d'Argent: Une Introduction*, Grenoble: Presses Universitaires de Grenoble.

de Brunhoff, Suzanne 2005, 'Marx's Contribution to the Search for a Theory of Money', in *Marx's Theory of Money: Modern Appraisals*, edited by Fred Moseley, New York: Palgrave Macmillan.

de Brunhoff, Suzanne and Duncan K. Foley 2007, 'Karl Marx's theory of money and credit', in *A Handbook of Alternative Monetary Economics*, edited by Philip Arestis and Malcolm Sawyer, Cheltenham: Edward Elgar Publishing.

De'Ath, Amy 2018, 'Reproduction', in *The Bloomsbury Companion to Marx*, edited by Jeff Diamanti et al., London: Bloomsbury Publishing.

Dinerstein, Ana Cecilia, Alfonso García Vela, Edith González and John Holloway (eds.) 2019, *Open Marxism 4: Against a Closing World*, London: Pluto Press.

Durand, Cédric 2017, *Fictitious Capital: How Finance Is Appropriating Our Future*, translated by David Broder, London: Verso.

Eagleton, Terry 1997, *Marx and Freedom*, London: Weidenfeld & Nicolson.

Elson, Diane (ed.) 2015 [1979], *Value: The Representation of Labour in Capitalism*, London: Verso.

Engels, Friedrich and Karl Marx 1962, *Selected Works*, Volume II, Moscow: Foreign Languages Publishing House.

Engels, Friederich and Karl Marx 1975, *Karl Marx and Frederick Engels: Collected Works*, New York: International Publishers.

Engels, Friederich and Karl Marx 1976, *Collected Works: Marx and Engels 1845–47*, Volume 5, London: Lawrence &Wishart.

Farris, Sara R. 2011, 'Workerism's Inimical Incursions: On Mario Tronti's Weberism', *Historical Materialism*, 19, 3: 29–62.

Federici, Silvia 2012, *Revolution at Point Zero: Housework, Reproduction, and Feminist Struggle*, Oakland: PM Press.

Ferguson, Susan 2014, 'A Response to Meg Luxton's "Marxist Feminism and Anticapitalism"', *Studies in Political Economy*, 94, 1: 161–68.

Ferguson, Susan 2017, 'Children, Childhood and Capitalism: A Social Reproduction Perspective' in *Social Reproduction Theory: Remapping Class, Recentring Oppression*, edited by Tithi Bhattacharya, London: Pluto Press.

Ferguson, Susan 2019, *Women and Work: Feminism, Labour, and Social Reproduction*, London: Pluto Press.

Feuerbach, Ludwig 2012, *The Fiery Brook: Selected Writings*, translated by Zawar Hanfi, London: Verso.

Fichte, Johann Gottlieb 1993, *Foundations of Transcendental Philosophy (Wissenschaftslehre) nova method (1796–99)*, Ithaca: Cornell University Press.

Filippini, Michele and Emilio Macchia 2012, *Leaping Forward: Mario Tronti and the History of Political Workerism*, Maastricht: Jan van Eyck Academie / Centro per la riforma dello Stato.

Firestone, Shulamith 2015 [1970], *The Dialectic of Sex: The Case for Feminist Revolution*, London: Verso.

Foley, Duncan K. 1986, *Understanding Capital: Marx's Economic Theory*, Cambridge, Mass.: Harvard University Press.

Foster, John Bellamy 2000, *Marx's Ecology: Materialism and Nature*, New York: Monthly Review Press.

Foucault, Michel 1991, *The Foucault Effect: Studies in Governmentality with Two Lectures by and an Interview with Michel Foucault*, edited by Graham Burchell et al., Chicago: University of Chicago Press.

Fraser, Nancy 2013, *Fortunes of Feminism: From State-Managed Capitalism to Neoliberal Crisis*, London: Verso.
Fraser, Nancy 2014, 'Beyond Marx's Hidden Abode', *New Left Review*, 86: 55–72.
Fraser, Nancy 2016, 'Contradictions of Capital and Care', *New Left Review*, 100: 99–117.
Fraser, Nancy 2017, 'Crisis of Care? On the Social-Reproductive Contradictions of Caontemporary Capitalism' in *Social Reproduction Theory: Remapping Class, Recentring Oppression*, edited by Bhattacharya, Tithi, London: Pluto Press.
Fraser, Nancy 2021, 'The Climates of Capital', *New Left Review*, 127: 94–127.
Frass, Carl 1858, *Die Natur in der Wirtschaft*. Erschöpfung und. Ersatz. Westermann's. Jahrbuch der illustrirten Deutschen Monatshefte, vol. 3.
Giménez, Martha E. 2019, *Marx, Women, and Capitalist Social Reproduction*, Leiden: Brill, 2018.
Gonzalez, Maya and Jeanne Neton 2014, 'The Logic of Gender: On the Separation of Spheres and the Process of Abjection', in *Contemporary Marxist Theory: A Reader*, edited by Andrew Pendakis et al., London and New York: Bloomsbury.
Graeber, David 2011, *Debt: The First 5,000 Years*, Brooklyn: Melville House Publishing.
Graziani, Augusto and Michel Vale 1997, 'Let's Rehabilitate the Theory of Value,' *International Journal of Political Economy*, 27, 2: 21–5.
Graziani, Augusto 1989, *The Theory of the Monetary Circuit*, London: Thames Polytechnic, School of Social Sciences.
Graziani, Augusto 1997, 'The Marxist Theory of Money', *International Journal of Political Economy*, 27, 2: 26–50.
Hardt, Michael and Antonio Negri 1994, *Labor of Dionysus: A Critique of the State-Form*, Minnesota: University of Minnesota Press.
Harvey, David 1982, *The Limits to Capital*, Oxford: Blackwell Publishers.
Hegel, Georg Wilhelm Friedrich 1977 [1807], *Hegel's Phenomenology of Spirit*, translated by A. V. Miller, Oxford: Oxford University Press.
Hegel, Georg Wilhelm Friedrich 2010 [1812–16], *The Science of Logic*, translated by George Di Giovanni, Cambridge: Cambridge University Press.
Hegel, Georg Wilhelm Friedrich 2008 [1820], *Outlines of the Philosophy of Right*, translated by T. M. Knox, Oxford: Oxford University Press.
Hegel, Georg Wilhelm Friedrich 1957 [1821], *Philosophy of Right*, translated by T. M. Knox, Oxford: Oxford University Press.
Hegel, Georg Wilhelm Friedrich 1991 [1830], *The Encyclopaedia Logic: Part I of the Encyclopaedia of Philosophical Sciences with the Zusätze*, translated by T. F. Geraets et al., Indianapolis: Hackett Publishing Company Inc.
Hegel, Georg Wilhelm Friedrich 1956 [1837], *The Philosophy of World History*, translated by J. Sibree, Mineola: Dover.
Hegel, Georg Wilhelm Friedrich 1999, *Political Writings*, edited by Laurence Dickey and H.B. Nisbet, translated by H. B Nisbet, Cambridge: Cambridge University Press.

Hegel, Georg Wilhelm Friedrich 2006–2009, *Lectures on the History of Philosophy: 1825–26*, 3 Volumes, edited by Robert F. Brown, Oxford: Oxford University Press.

Hegel, Georg Wilhelm Friedrich 2012, *Lectures on Natural Right and Political Science: The First Philosophy of Right*, translated by J. Michael Stewart and Peter C. Hodgson, Oxford: Oxford University Press.

Heinrich, Michael 2012, *An Introduction to the Three Volumes of Karl Marx's Capital*, translated by Alex Locascio, New York: Monthly Review Press.

Heller, Agnes 2018, *The Theory of Need in Marx*, London: Verso Books.

Hilferding, Rudolf 2006 [1981], *Finance Capital: A study of the latest phase of capitalist development*, translated by Morris Watnick and Sam Gordon, London: Routledge.

Hope, Wayne 2011, 'Crisis of temporalities: Global capitalism after the 2007–08 financial collapse', *Time and Society*, 20, 1: 94–118.

Howie, Gillian 2009, 'Breaking Waves: Feminism and Marxism Revisited', in *Karl Marx and Contemporary Philosophy*, edited by Andrew Chitty and Martin McIvor, London: Palgrave MacMillan.

Hyppolite, Jean 1969, *Studies on Marx and Hegel*, New York: Basic Books.

Inwood, Michael (ed.) 1992, *A Hegel Dictionary*, Oxford: Blackwell Publishers.

Jameson, Fredric 2011, *Representing Capital: A Reading of Volume One*, London: Verso.

Kant, Immanuel 1965 [1781], *Immanuel Kant's Critique of Pure Reason*, translated by Norman K. Smith, New York: St Martin's Press.

Kant, Immanuel 2002 [1788], *Critique of Practical Reason*, translated by Werner S. Pluhar, Indianapolis: Hackett Publishing Company Inc.

Kant, Immanuel 2000 [1790], *Critique of the Power of Judgment*, translated by Eric Matthews, Cambridge: Cambridge University Press.

King, J. E., and M. C. Howard 1989–1992, *A History of Marxian Economics*, 2 Volumes, Basingstoke: Macmillan.

Kolakowski, Leszek 1978, *Main Currents of Marxism: Its Rise, Growth, and Dissolution*, 3 Volumes, translated by P. S. Falla, Oxford: Clarendon Press.

Korsch, Karl 1970 [1923], *Marxism and Philosophy*, translated by Fred Halliday, New York: Monthly Review Press.

Kuruma, Samezō 1957, *Theory of the Value-Form & Theory of the Exchange Process*, translated by E. Michael Schauerte, Tokyo: Iwanami Shoten.

Lange, Elena Louisa 2014, 'Failed Abstraction – The Problem of Uno Kōzō's Reading of Marx's Theory of the Value Form', *Historical Materialism*, 22, 1: 3–33.

Lange, Elena Louisa 2016, 'The Critique of Political Economy and the "New Dialectic": Marx, Hegel and the Problem of Christopher J. Arthur's "Homology Thesis"', *Crisis & Critique*, 3,3: 235–72.

Lapavitsas, Costas 2013, *Profiting Without Production: How Finance Exploits Us All*, London: Verso.

Lazzarato, Maurizio 2012, *The Making of the Indebted Man: An Essay on the Neoliberal Condition*, translated by Joshua David Jordan, New York: Semiotext(e).

Lazzarato, Maurizio 2015, *Governing by Debt*, translated by Joshua David Jordan, New York: Semiotext(e).

Lebowitz, Michael A. 2009, *Following Marx: Method, Critique and Crisis*, Leiden: Brill, 2009.

Liebig, Justus von 1843, *Familiar letters on chemistry and its relation to commerce, physiology and agriculture*, New York: J. Winchester, New World Press.

Lenin, Vladimir Ilyich 1972, *Collected Works*, Volume 38, Moscow: Philosophical Notebooks.

Lucarelli, Stefano 2013, 'The 1973–1978 workgroup on money of the journal "Primo Maggio": an example of pluralist critique of political economy', *The International Journal of Pluralism and Economics Education*, 4, 1: 30–50.

Lukács, György 2010 [1908], *Soul and Form*, translated by Anna Bostock, New York: Columbia University Press.

Lukács, György 1971 [1923], *History and Class Consciousness*, translated by Rodney Livingstone, London: Merlin Press.

Luxemburg, Rosa 2003 [1913], *The Accumulation of Capital*, translated by Agnes Schwarzschild, Milton Park: Routledge Classics.

Luxton, Meg 2006 "Feminist Political Economy in Canada and the Politics of Social Reproduction," in *Social Reproduction: Feminist Political Economy Challenges Neoliberalism*, Kate Bezanson and Meg Luxton (ed.), Montréal: McGill-Queen's University Press.

Maksakovsky, Pavel V. 2004, *The Capitalist Cycle: An Essay on the Marxist Theory of the Cycle*, Leiden: Brill.

Malm, Andreas 2016, *Fossil Capital: The Rise of Steam Power and the Roots of Global Warming*, London: Verso.

Marazzi, Christian 2015, 'Money and Financial Capital', *Theory, Culture & Society*, 32, 7–8: 39–50.

Marx, Karl 1847, *Wage Labour and Capital*, available at: https://www.marxists.org/archive/marx/works/1847/wage-labour/.

Marx, Karl 1970 [1859], *A Contribution to the Critique of Political Economy*, New York: New World Paperbacks.

Marx, Karl 1863, *Theories of Surplus Value*, available at: https://www.marxists.org/archive/marx/works/1863/theories-surplus-value/.

Marx, Karl 1867, *Das Kapital: Kritik der politischen Oekonomie*, Hamburg: Verlag von Otto Meissner.

Marx, Karl 1990 [1867], *Capital: A Critique of Political Economy, Volume I: The Process of Production of Capital*, translated by Ben Fowkes, New York: Penguin Books.

Marx, Karl 1992 [1884], *Capital: A Critique of Political Economy, Volume II: The Pro-*

cess of Circulation of Capital, translated by David Fernbach, New York: Penguin Books.
Marx, Karl 1991 [1895], Capital: A Critique of Political Economy, Volume III: The Process of Capitalist Production as a Whole, translated by David Fernbach, New York: Penguin Books.
Marx, Karl 1973 [1939], Grundrisse: Foundations of the Critique of Political Economy (Rough Draft), translated by Martin Nicolaus, Harmondsworth: Penguin Books.
Marx, Karl 2008, 'Werke – Band 23: Das Kapital – Band 1', in Marx-Engels-Werke (MEW), Berlin: Karl-Dietz-Verlag.
Marx, Karl 2015, Marx's Economic Manuscript of 1864–1865, edited by Fred Moseley, translated by Ben Fowkes, Leiden: Brill.
Marx, Karl 2019, The Political Writings, London: Verso.
Mattick, Paul 2019, Theory as Critique: Essays on Capital, Chicago: Haymarket Books.
Mau, Søren 2021, 'The Mute Compulsion of Economic Relations: Towards a Marxist Theory of the Abstract and Impersonal Power of Capital', Historical Materialism, 29, 3: 3–32.
McNally, David 2009, 'From Financial Crisis to World-Slump: Accumulation, Financialisation, and the Global Slowdown', Historical Materialism, 17, 2: 35–83.
Meaney, Mark E. 2002, Capital as Organic Unity: The Role of Hegel's Science of Logic in Marx's Grundrisse, London: Springer.
Mészáros, István 1970, Marx's Theory of Alienation, London: Merlin Press.
Mohun, Simon (ed.) 1994, Debates in Value Theory, Basingstoke: Palgrave Macmillan.
Moore, Jason 2015, Capitalism in the Web of Life: Ecology and the Accumulation of Capital, London: Verso.
Moseley, Fred (ed.) 2005, Marx's Theory of Money: Modern Appraisals, London: Palgrave Macmillan.
Moseley, Fred 2015, Money and Totality: A Macro-Monetary Interpretation of Marx's Logic in Capital and the End of the 'Transformation Problem', Leiden: Brill.
Moseley, Fred and Tony Smith (eds.) 2015, Marx's Capital and Hegel's Logic: A Reexamination, Chicago: Haymarket Books.
Negri, Antonio 1988, Revolution Retrieved: Selected Writings on Marx, Keynes, Capitalist Crisis and New Social Subjects 1967–1983, London: Red Notes.
Negri, Antonio 1991, Marx Beyond Marx: Lessons on the Grundrisse, translated by Harry Cleaver et al., New York and London: Autonomedia / Pluto.
Negri, Antonio 2003, Time for Revolution, translated by Matteo Mandarini, London and New York: Continuum.
Nelson, Anitra 1999, Marx's Concept of Money, New York: Routledge.
Ng, Karen 2020, Hegel's Concept of Life: Self-Consciousness, Freedom, Logic, Oxford: Oxford University Press.
Nigro, Roberto 2018, 'Workerism,' Krisis: Journal for Contemporary Philosophy, 2: 173–76.

Osborne, Peter 1995, *The Politics of Time: Modernity and Avant-Garde*, London: Verso.
Osborne, Peter 2004, 'The reproach of abstraction', *Radical Philosophy*, 127: 21–8.
Osborne, Peter 2005, *How to Read Marx*, London: Granta Books.
Osborne, Peter 2008, 'Marx and the philosophy of time', *Radical Philosophy*, 47: 15–22.
Osborne, Peter 2010, 'A sudden topicality: Marx, Nietzsche and the politics of crisis', *Radical Philosophy*, 160: 19–26.
Osborne, Peter, Éric Alliez and Eric-John Russell (eds.) 2019, *Capitalism; Concept, Idea, Image – Aspects of Marx's Capital Today*, London: CRMEP Books.
Osborne, Peter (ed.) 2020, 'Thinking Art: Materialisms, Labours, Forms', London: CRMEP Books.
Pashukanis, Evgeny B 1989 [1924], *Law & Marxism*, London: Pluto Publishing Ltd.
Piketty, Thomas 2014, *Capital in the Twenty-First Century*, translated by Arthur Goldhammer, Cambridge, Mass. and London: The Belknap Press of Harvard University Press.
Pippin, Robert B. 2018, *Hegel's Realm of Shadows: Logic as Metaphysics in "The Science of Logic"*, Chicago: University of Chicago Press.
Postone, Moishe 1993, *Time, Labour and Social Domination: A Reinterpretation of Marx's Critical Theory*, Cambridge: Cambridge University Press.
Postone, Moishe 2007, *Critical Theory, Philosophy, and History*, Presentation: Congrès Marx International V – Section Philosophie – Capital – Paris-Sorbonne et Nanterre.
Postone, Moishe 2016, *Capitalism, Temporality and The Crisis of Labour*, Presentation: American Academy Berlin.
Reichelt, Helmut 2005, 'Social Reality as Appearance: Some Notes on Marx's Conception of Reality', in *Human Dignity: Social Autonomy and the Critique of Capitalism*, edited by Werner Bonefeld and Kosmos Psychopedis, Aldershot: Ashgate.
Roberts, Michael 2016, *The Long Depression: Marxism and the Global Crisis of Capitalism*, Chicago: Haymarket Books.
Roberts, William Clare 2017, *Marx's Inferno: The Political Theory of Capital*, Princeton: Princeton University Press.
Robinson, Cedric J. 2005 [1983], *Black Marxism: The Making of the Black Radical Tradition*, Chapel Hill: University of North Carolina Press.
Rose, Gillian 1981, *Hegel Contra Sociology*, London: Verso.
Rubin, I.I. 2016 [1923], *Essays on Marx's Theory of Value*, Delhi: Aakar Books.
Saad-Filho, Alfredo 2002, *The Value of Marx: Political Economy for Contemporary Capitalism*, London: Routledge.
Saito, Kohei 2017, *Karl Marx's Ecosocialism: Capital, Nature, and the Unfinished Critique of Political Economy*, New York: Monthly Review Press.
Sargent, Lydia (ed.) 1981, *Women and Revolution: A Discussion of the Unhappy Marriage of Marxism and Feminism*, Boston: South End Press.

Schelling, F. W. J. von 1988 [1797], *Ideas for a Philosophy of Nature*, translated by Errol E. Harris and Peter Heath, Cambridge: Cambridge University Press.

Schelling, F. W. J. von 2003 [1809], *Philosophical Inquiries Into the Nature of Human Freedom*, translated by Jeff Love and Johannes Schmidt, London: Open Court.

Schmidt, Alfred 2014 [1962], *The Concept of Nature in Marx*, London: Verso.

Sekine, Thomas 2019, *The Dialectic of Capital*, 2 Volumes, Leiden: Brill.

Sewell, William H. 2008, 'The Temporalities of Capitalism', *Socio-Economic Review*, 6, 3: 517–37.

Simmel, Georg 1990 [1900], *Philosophy of Money*, London: Routledge.

Simon, Lawrence H. (ed.) 1994, *Karl Marx: Selected Writings*, Indianapolis and Cambridge: Hackett Publishing Company, Inc.

Sohn-Rethel, Alfred 1978, *Intellectual and Manual Labor: A Critique of Epistemology*, London: Macmillan.

Spinoza, Benedictus de 1996 [1677], *Ethics*, translated by Edwin Curley, London: Penguin Classics.

Sutherland, Zöe and Marina Vishmidt 2020, 'Social Reproduction: New Questions for the Gender, Affect and Substance of Value', in *The New Feminist Literary Studies*, edited by Jennifer Cooke, Cambridge: Cambridge University Press.

Tomba, Massimiliano 2009, 'Historical Temporalities of Capital: An Anti-Historicist Perspective', *Historical Materialism*, 17, 4: 44–65.

Tomba, Massimiliano 2013, *Marx's Temporalities*, Leiden: Brill.

Tombazos, Stavros 2015, *Time in Marx: The Categories of Time in Marx's Capital*, Chicago: Haymarket Books.

Trigg, Andrew B. 2006, *Marxian Reproduction Schema: Money and aggregate demand in a capitalist economy*, Milton Park: Routledge.

Tronti, Mario 1971, *Workers and Capital*, translated by David Broder, London: Verso.

van der Linden, Marcel and Karl Heinz Roth (eds.) 2014, *Beyond Marx: Theorising the Global Labour Relations of the Twenty-First Century*, Leiden: Brill.

Vishmidt, Marina 2012, 'Counter(Re-)Productive Labour', *Auto Italia Blog*, available at: https://autoitaliasoutheast.org/blog/counter-re-productive-labour/.

Vogel, Lise 2014, *Marxism and the Oppression of Women: Toward a Unitary Theory*, Chicago: Haymarket Books.

Wood, Ellen Meiksins 1999, *The Origin of Capitalism*, London: Verso Books.

Wood, Ellen Meiksins 2016, *Democracy Against Capitalism: Renewing Historical Materialism*, London: Verso Books.

Wright, Steve 2014, 'Revolution from Above? Money and Class-Composition in Italian Operaismo', in *Beyond Marx: Theorising the Global Labour Relations of the Twenty-First Century*, edited by Marcel van der Linden and Karl Heinz, Leiden: Brill.

Zacarés, Javier Moreno 2021, 'Euphoria of the Rentier?', *New Left Review*, 129: 47–67.

# Index

Absolute   137, 146, 160, 161, 167
Arruzza, Cinzia   6
Arthur, Christopher, J.   116, 121, 130–134, 153, 158, 160–164, 173
Automatic Subject   3, 14, 99, 138, 154n28, 170, 182

Backhaus, Hans-Georg   1n2, 31n5, 73n5, 82n34
Balibar, Étienne   1, 95, 97, 98, 101, 179
Begriff   see Concept
Being   78–79, 151–154, 157–159, 162, 164, 166–170, 173, 178
Bellofiore, Riccardo   46n33, 64n90, 66n96, 68n104
Bhandar, Brenna and Alberto Toscano   13–15
Bhattacharya, Tithi   6
Bretton Woods   40n11

Campbell, Martha   134
Capital
   Accumulation   3, 18, 21, 23, 26, 58, 68, 116, 120, 125, 131, 184
Banking   69, 109n13, 110n13, 126, 129, 131–132
   Constant   105, 117, 120, 123–124, 133
   Fictitious   3–5, 13–18, 20, 22–27, 30–32, 35, 47, 54–57, 84, 88–89, 94, 103, 106, 111–112, 129, 184
      Finance   2, 6, 11, 25, 134, 186–187
      Interest-bearing   13, 16n13, 69, 72, 84–89, 106, 129, 130–131, 134
      Logic   6, 36, 52, 57–58
   Variable   105, 110–111, 116, 120, 123–125, 133
Chakrabarty, Dipesh   26
Character Mask   29n47, 80, 91, 96, 101, 104n11
*Classe Operaia*   43–44
Collins, Patricia Hill   28n45
Concept   126–127, 133, 136–137, 146–147, 149–150, 152–154, 157–162, 165–171, 173, 183
Credit   13, 15, 17–21, 33–38, 47, 51–54, 56, 57, 59, 62, 70, 87, 89, 106, 112, 135, 161n42
   System   15, 18, 19, 54, 68, 88, 128–130, 134, 135
*Credit Mobilier*   47

Creditor-debtor relation   13, 17, 19, 20, 23, 31, 51, 103, 138
Crenshaw, Kimberle   28n45

de Brunhoff, Suzanne   3n5, 6, 7, 18–20, 32–38, 39n9, 51–61, 66–70, 88, 89, 135
De'Ath, Amy   9, 117n26, 133
Debt   13–15, 19–21, 184
   State   53
Domination   12, 143, 184
   Personal/Interpersonal   5, 14, 16, 51, 90, 143n9
   Impersonal   5n8, 12, 14–16, 93n61
   Abstract   12–14, 28, 98, 100
   Extra-economic   32n52, 99, 100–103
Durand, Cédric   6

Ecology   1, 4, 11, 140, 142, 146, 148, 172, 174, 176, 180–181, 186
Essence   39, 56, 72–74, 76–80, 26, 137, 151–153, 157, 159, 162, 164, 166–167, 173, 176–177, 179–183

Farris, Sara R.   43
Federici, Silvia   106n4, 139n90
Ferguson, Susan   6, 28, 30, 142
Fetish Character   8, 10, 12, 66, 68–88, 94–95, 102, 113, 125–126, 136, 150, 157, 162
Fetishism   3n6, 8, 15, 21, 35, 71, 73, 75, 82, 83, 90, 96–97, 105
Feuerbach, Ludwig   139, 149n18, 172, 175, 176n74, 177–183
Financialisation   2, 5, 10, 13, 16, 26, 36, 38n8, 45–46, 111, 139, 142, 184–185
Firestone, Shulamith   15
Foley, Duncan K.   54, 61, 65n92
Foucault, Michel   31, 47n38
Fraser, Nancy   6, 32n52, 142, 145, 146

Geert Reuten   131, 132, 134
Giménez, Martha   6
Graziani, Augusto   46, 60, 68n104

Harvey, David   15, 23, 27n43
Hegel, Georg Wilhelm Friedrich   9, 63, 72, 76–79, 96, 97, 100, 110, 112, 126,

Hegel, Georg Wilhelm Friedrich *cont.*   137, 139, 141, 146–171, 173, 174, 182, 183, 184n90
Heinrich, Michael   1n1, 14, 22, 23, 73n5, 76n14, 81, 82n34, 83n38
Heller, Agnes   137n86

Immanent Externality   7, 61, 70, 164, 174
Interruption   3, 5, 8, 104, 105, 107, 108, 111, 114, 119, 122, 127–129, 134, 138, 147, 160, 161
Inversion   4n6, 54, 71, 73–76, 79–81, 88, 91, 116, 125–126, 136, 150

Juridical Mask   8, 29n47, 93–96, 99, 101–103

Kant, Immanuel   63
Korsch, Karl   102

Labour
  Abstract   3n6, 5n8, 12, 21, 23, 25, 33, 56, 62, 65, 67, 69n104, 73–76, 78–81, 86–88, 109–112, 114, 116, 125–126, 135–138, 143, 156–158, 162–164, 168, 171n64, 173n68
  Concrete   12, 74, 76
Power   5, 12, 14–15, 25, 27–31, 42, 48, 56–57, 64, 69n104, 79, 92, 98, 107n6, 120, 122, 140–142, 145, 147n16, 175–176, 184n90, 185
  Time   12, 18, 21, 24–26, 30, 53, 65–66, 72, 76, 79–80, 114, 116
  Living   11n29, 67, 109, 136, 148, 155, 157, 162–164, 173n68
Lange, Elena Louisa   8, 73n5, 74, 75, 76n13, 126n50–51, 136, 150n19, 158
Lapavitsas, Costas   5, 15n13, 25n38, 38n8
Lapo Berti   36, 38, 45, 47, 48–52, 54–56
Liebig, Justus von   177n75
Life
  Of capital   11, 83, 93, 94, 109, 118, 120, 129, 132, 137, 147–150, 154–156, 159, 160, 162, 164, 174, 175, 182, 183, 186
  Human   3, 4, 11, 37, 80, 83, 84, 89, 107, 109, 110–115, 120, 122, 129, 136–142, 144, 147, 150, 162, 164, 172, 173, 177n76, 179, 182, 183, 185–187
  Natural   187, 148
  Concrete   4, 9, 10, 104, 107, 139, 147, 148, 153, 172, 174, 176, 183, 184, 186
  Abstract   174, 175

Lukács, György   113
Luxemburg, Rosa   106n4
Luxton, Meg   142

Marina Vishmidt   141n2
Mau, Søren   14
Meaney, Mark E.   10, 110, 127n53, 152, 159n39, 168n59
Metabolism   144, 154, 156, 172, 173, 175n72, 177n75
Money
  Credit   2, 10, 13, 18–21, 30, 33–35, 37, 39, 47, 52–53, 56, 58, 62, 70, 88, 107, 111–112, 128–130, 134, 138, 142n7, 160, 173, 184
  As Money   1, 6–7, 33–34, 36–38, 51–52, 59, 69–70, 84, 86, 108, 142n7, 157, 160–161, 171, 173
Moseley, Fred   9, 60, 65, 67n102, 117
Mystification   27, 71–72, 82–83, 87, 90, 101

Nature   37, 82–83, 93n60, 94n63, 104–105, 107, 110–111, 115, 139–141, 144–149, 154, 156, 162, 164, 169, 172–173, 176n73, 176n74, 177, 180–183, 186–187
Negri, Antonio   39n10, 41n15, 48
Ng, Karen   9, 167, 168
Notion   see Concept

Objectivity   3, 73, 75, 78–79, 82, 116, 157, 160, 169, 171, 179
Organic Unity   152, 162

Pashukanis, Evgeny   9, 72, 90, 93n61, 97, 98, 100, 102
Person   8, 14, 16, 21–22, 24, 29n47, 72–74, 80, 81n32, 89–103
  Personification   3, 13n5, 71, 73, 80–81, 87, 91, 95–96, 98–101, 155n29, 170
Political Subject   36–37, 39n10, 40, 42–43, 58–59
  Post-Kantian   139
Postone, Moishe   14, 25, 26, 62, 63, 184n90
*Primo Maggio*   38, 45–48, 52, 58, 68

Reification   35, 71–75, 80–81, 83, 90, 94–95, 100–101, 110, 112, 151n19, 163, 173, 182
Reproduction
  Capitalist   4, 10, 105, 113, 129, 135, 139, 186–187

INDEX

Concrete   105, 140, 143, 150, 165, 172, 174
Expanded   8, 109n13, 122, 126, 130–136, 138, 157, 159, 165
Schemas of   8, 35, 106, 109n13, 117n26, 176
Social   2, 4–6, 8–9, 11, 14–16, 28–31, 32n52, 37, 61, 68, 84, 89, 103, 106–107, 114, 120, 122, 125, 134, 138, 139–142, 144–146, 148, 156, 174, 184–187

Rubin, I.I.   7, 71n1, 73n5, 75n10, 95, 96

Saad-Filho, Alfredo   55n74
Saito, Kohei   1n1, 176n74, 177n75, 179n81, 180n83
Schmidt, Alfred   176n74
Species-being   177
Subjection   2–3, 4n6, 7–8, 10, 16–17, 20–21, 24, 32, 35–38, 58–59, 69n105, 70–71, 81, 82, 84, 87, 89, 94, 99, 102, 107, 109–110, 113–115, 122, 136, 153, 170, 172, 186
Subjectivity   30, 39, 42–43, 58–59, 73, 79–80, 97, 101, 113, 116, 157, 159–160, 162, 179, 182

The Trinity Formula   82, 83, 100, 101
Time
 Free   27n44, 114
 Of Life   104, 111, 113
 Of Capital   111, 113, 114, 139
 Turnover   108, 111, 121, 122, 128, 133
 Circulation   110, 121, 133, 138
 Of social practice   21, 113
 Of organic being   114
 Of production   23, 79, 105
Tomba, Massimiliano   8, 25, 26, 114, 118n29, 139n89
Tombazos, Stavros   8, 76–80, 118n29, 128n59, 133n75, 134n76, 151n20, 152, 157

Valorisation   3n6, 4–5, 5n8, 8, 12–13, 15, 17, 21, 24, 24n35, 26–27, 30–31, 34, 59, 69n104, 73, 75, 82–84, 99, 101, 105–106, 109–111, 115–118, 120–121, 123, 125–129, 132, 135–138, 151, 155–157, 159, 161–162, 170
*Value-in-itself*   162, 163, 166
Vogel, Lise   29, 140n1, 174, 184n90

Wood, Ellen Meiksins   143n9

Zacarés, Javier Moreno   94